STABILITY AND COMPLEXITY IN
Model
Ecosystems

·GRANDIBVS EXIGVI SVNT PISCES PISCIBVS ESCA·

Siet sone dit hebic ick zeer langhe gh'weten dat die groote vissen de cleyne

# STABILITY AND COMPLEXITY IN

# Model

# Ecosystems

ROBERT M. MAY

PRINCETON, NEW JERSEY

PRINCETON UNIVERSITY PRESS

MONOGRAPHS IN POPULATION BIOLOGY

EDITED BY ROBERT H. MacARTHUR

LC Card: 72-9948
ISBN: 0-691-08125-5 (hardcover edn.)
ISBN: 0-691-08130-5 (paperback edn.)

This book has been composed in Linofilm Baskerville.

Printed in the United States of America.

# Preface

This book surveys a variety of theoretical models, all bearing on aspects of population stability in biological communities of interacting species. Some of the broader themes are the relation between stability and complexity in general multispecies models; the relation between stability in randomly fluctuating environments as opposed to deterministic ones; and the way environmental fluctuations are liable to put a limit to niche overlap, a limit to similarity, among competing species in the real world. Minor themes include the way nonlinearities can produce stable limit cycle oscillations in real ecosystems; the role played by time-delays in feedback mechanisms, and the way that addition of extra trophic levels can stabilize them; the relation between stability within one trophic level and total web stability; and why strong predator-prey links may be more common in nature than strong symbiotic links. The survey is neither impersonal nor encyclopaedic, but rather is an idiosyncratic reflection of my own interests.

This work seeks to gain general ecological insights with the help of general mathematical models. That is to say, the models aim not at realism in detail, but rather at providing mathematical metaphors for broad classes of phenomena. Such models can be useful in suggesting interesting experiments or data collecting enterprises, or just in sharpening discussion. The book is primarily directed at the field and laboratory ecologist, and the text is hopefully accessible to people with minimal mathematical training. (In cases where the mathematical technicalities are likely to be of some general interest to theoretical biologists, they are set out in self-contained appendices.)

I am deeply indebted to many people for their patience and guidance. As a newcomer to ecology, I have been struck by the attitude of constructive interest in others' work which seems to prevail among ecologists. The competition and predation which characterize many other disciplines seem relatively absent, possibly because the field has not yet reached (or exceeded) its natural carrying capacity. My background is in theoretical physics, and I am at least aware of the danger that my interests are liable to be animated too much by elegance and too little by common sense. It is for the reader to judge whether I have benefited from that awareness.

Much of the material for this book was assembled while I was a Visiting Scientist at the Culham plasma physics laboratory and an Honorary Member of Magdalen College, Oxford, and later a Visiting Member at the Institute for Advanced Study, Princeton. I thank the people at these places for their kind hospitality. Professor H. Messel, Director of the Science Foundation for Physics within the University of Sydney, has generously maintained his verbal and financial encouragement whilst I strayed from his fold.

The number of people whose comments have helped to form this book is too large to list. A most incomplete catalogue includes L. C. Birch (who started it all), J. H. Connell, F. J. Dyson, N. G. Hairston, M. P. Hassell, H. S. Horn, S. P. Hubble, E. Leigh, S. Levin, R. Levins, R. C. Lewontin, M. Lloyd, J. Maynard Smith, W. W. Murdoch, D. Pimentel, J. Roughgarden, T. W. Schoener, R. O. Slatyer, L. B. Slobodkin, T. R. E. Southwood, and K. E. F. Watt. My gratitude, nonetheless sincere for its conventionality, is also due to Maria Dunlop, who typed the manuscript, to Ross McMurtrie and Brian Martin, who helped with the work, and to my wife and daughter for their interest.

Above all, I thank Robert MacArthur, without whose stimulating encouragement the book would not have been written.

Sydney University, 1972                                    R. M. M.

# Contents

STABILITY AND COMPLEXITY IN
# Model
# Ecosystems

# CHAPTER ONE

# Introduction

This book contains four loosely connected main themes, which are developed, one by one, in Chapters 3, 4, 5 and 6.

1. Chapter 3 draws together various lines of argument to suggest that, in general mathematical models of multispecies communities, increasing complexity tends to beget diminished stability. In pursuit of this generalization, we consider the stability character, first, of a particular class of multispecies predator-prey models (being rather dismissive of certain recent models with very special symmetry properties); second, of large, complex ecosystem models in which the trophic web links are assembled at random; and, third, of models in which we know only the topological structure of the trophic web, that is, only the signs of the interactions between the various species. (This third kind of approach is conventionally called "qualitative stability theory.") A miscellany of other arguments are also touched upon. Roughly speaking, we here take complexity to be measured by the number and nature of the individual links in the food web, and stability by the tendency for population perturbations to damp out, returning the system to some persistent configuration.

These results caution against any simple belief that increasing population stability is an automatic *mathematical* consequence of increasing multispecies complexity. That stability may usually go with complexity in the natural world, but not usually in general mathematical models, is not really paradoxical. In nature we deal not with arbitrary complex systems, but rather with ones selected by a long and intricate process. The emergent moral is that theoretical work should not try to prove any general theorem that "complexity implies stability," but instead should focus on elucidating the very special sorts of complexity, the singular strategies, which may promote such mathematically atypical stability.

In passing, we take up the question of the relation between stability in any one trophic level by itself and stability of the total trophic web, and show how a synthesis may be achieved among the diversity of views which have been expressed on this subject. Also, from the discussion of qualitative stability theory, it emerges that while predator-prey bonds in the food web tend to have a stabilizing influence, symbiosis or mutualism tends to be destabilizing. This suggests that stability considerations may play a part in explaining why symbiotic links between species are relatively uncommon in many natural ecosystems.

2. Chapter 4 deals with some interesting features of more detailed and realistic models for the dynamics of populations in communities with one, two, or three species.

A full nonlinear analysis of these models can yield stable limit cycle solutions, in which the populations oscillate up and down in a stable periodic manner, between maxima and minima determined wholly by the intrinsic parameters of the model. It is shown that essentially all 1 predator–1 prey models in the literature admit naturally of a range of stable limit cycle solutions, particularly in cases where self-limitation effects in the prey birth rates are relatively weak. On another tack, single species models incorporating a stabilizing density-dependent feedback term, with a time-delay, can also exhibit stable limit cycle behavior if the time-delay is too long. The time-delayed logistic model first studied by Hutchinson (1948) provides the standard example. These limit cycle ideas are developed with reference to such varied natural phenomena as the Hudson Bay lynx and hare (Figure 4.4) and Nicholson's (1954) blow-flies, which latter example is studied in some detail.

The role played by time-delays in the community inter-actions is further pursued, bearing in mind the engineer's axiom that, if a potentially stabilizing feedback loop is ap-plied with a time lag that is long compared with the natural time scale of the system, it will in fact act as a destabilizing element. This idea is developed for vegetation-herbivore and vegetation-herbivore-carnivore systems in which the stabilizing resource-limitation effect operates on the herbi-vore population with a time lag. Under certain conditions, which are commonly met in nature, the vegetation-herbi-vore-carnivore system is stable (with population fluctuations being damped out), while the vegetation-herbivore system with no predators is unstable. This model, which can be supported with some observational data, suggests inter alia that herbivore population numbers may often be set neither by predators alone nor by vegetation alone, but by an explicit interplay between both effects.

3. Chapter 5 treats the relation between the dynamics of population models in which all the environmental param-eters are strictly deterministic and the corresponding, more realistic models with random environmental fluctua-tions. On this basis we discuss the connection between the deterministic, mechanical usage of the term "stability" (defined as the propensity, following a perturbation, to return to the deterministic equilibrium point), and that usage which associates stability or instability with the de-gree of population fluctuation in a stochastic environment.

Although such random variations in the environment can introduce qualitatively new features into the model, it can often be that the main results of the deterministic analysis survive in these more realistic models. This sup-plies a retrospective justification for the attention given to deterministic models in Chapters 3 and 4.

4. Chapter 6 takes up the question of niche overlap, or limiting similarity, among competing species.

The problem is studied for a class of model biological communities in which several species compete on a one-dimensional continuum of resources, for example food size or vertical habitat. The resource spectrum is taken to have an element of random fluctuation. Within this framework, there emerges a robust mathematical result, namely that there is an effective limit to niche overlap consistent with long-term stability, and that this limit is insensitive to the degree of environmental fluctuation, unless it be very severe.

This conclusion marches with Hutchinson's (1959) classic observation that, in a variety of circumstances, including both vertebrate and invertebrate forms, character displacement among sympatric species leads to sequences in which each species is roughly twice as massive as the next; that is, linear dimensions as measured by bills or mandibles are in the ratio around 1.2 to 1.4. MacArthur's (1971b, 1972) more recent and more quantitative reviews of data culled from circumstances to which the one-dimensional competition theory seems roughly applicable (mainly congeneric birds, sorting out by food size or by vertical habitat height) provides further substantiation of this limit to niche overlap.

It is to be emphasized that the models of Chapter 6 are subject to two severe restrictions. First, they treat competition on a one-dimensional resource continuum. More generally, particularly for insects and plants, the niche will be multidimensional, with many relevant resource dimensions all intertwined. Second, the model is essentially confined to competition within one trophic level. While this is liable to be applicable to the higher animals near the top of the trophic web, it can well be in the lower trophic levels that competition is less important than the pressures

6

of predation, or other effects. These restrictions have been kept in mind by MacArthur and others, as is evidenced by the sorts of communities from which the data are gleaned. In developing a theory of niche overlap, it makes sense to begin with this restricted but relatively straightforward circumstance. One may hope to extend the theory by including more niche dimensions, and by embedding the set of competitors within a web, with predators above them.

The final Chapter 7 begins by looking again at our handful of pieces relating to the stability-complexity jigsaw puzzle. We conclude with cocktail party speculations as to some broad questions the theoretical ecology of the future may seek to answer.

So much for the ground covered in the book. The ground not covered should be emphasized.

Various aspects of the dynamics of population models are treated, but never with any long-term time dependence in the parameters of the models. There is no explicit evolution in our model ecosystems. The underlying genetic mechanisms are likewise not dealt with.

Furthermore, we consider isolated communities which are uniformly unvarying in space; time is the only independent variable. Spatial inhomogeneities, however, undoubtedly play a major role in many, if not most, real biological communities. The interplay between migration and extinction in a number of local populations in a spatially heterogeneous environment can have a stabilizing effect of the "not-putting-all-eggs-in-one-basket" kind. These points have been made in a general way, with reference to a variety of biological circumstances, by Levins (1969a, 1970a), Huffaker (1958), Den Boer (1968), MacArthur and Wilson (1967), Smith (1972), and others. The ideas have been illustrated in numerical simulations by Maynard Smith

(1971), St. Amant (1971), Roff (1972), and Reddingius and Den Boer (1970). One of the major theoretical treatments is due to Skellam (1951), who observed that, if one of two hypothetical plant species is superior in dispersal, it can persist, even if the other species would always out-compete it in a spatially static situation; this theme has been extended by Levins (1970a) and by Horn and MacArthur (1972).

Such spatial complications, interesting and frequently relevant though they surely are, have not been covered here. The excuse is that there remain worthwhile things to be done, before adding literally another dimension to the problem.

## THE MATHEMATICS IN THIS BOOK

Since this is a book about mathematical models of biological communities of interacting species, it is perforce cast in a mathematical mold. However, the pious aim is to communicate such insights as emerge from these models to field and laboratory ecologists. In the main part of the book, Chapters 3 through 7, the underlying assumptions which generate the various mathematical models are discussed in biological terms. With this groundwork laid, we usually go directly to set out the conclusions which emerge, again with emphasis on the biological morals to be drawn. The intervening mathematical jiggery-pokery, which carries us from initial assumptions to final results, is not dwelt upon in the text.

An extensive series of appendices makes good these lacunae in the text. The appendices are of two kinds (neither of which need be read). Some illustrate simple mathematical points with detailed examples. These are designed for the mathematically ill-equipped reader, who may like to see things spelt out very explicitly in a way which would distractingly clutter the main text for many readers. Other appendices are directed in quite the opposite sense, and elaborate some of the mathematical technicalities which enter into the detailed development of the models. Such appendices are largely self contained, and cover topics where the techniques may be of interest and use in other areas of mathematical biology (for example, the eigenvalue spectrum of certain classes of large matrices). In cases where the mathematical techniques are both recondite and narrowly specific to the problem at hand (as in much of Chapter 6), we give only the conclusion and a reference to the original literature.

The necessary minimal mathematical scaffolding for the subsequent chapters is erected in Chapter 2. This is tactically convenient, although it is perhaps a strategic error, in that such a very unbiological beginning may be off-putting.

Chapter 2 begins with a discussion of meanings which may be attached to "stability," and then outlines some formal methods of stability analysis for model ecosystems. Particular attention is paid to the "community matrix" (Levins, 1968a), an entity which, on the one hand, summarizes the biology of the community of interacting species near equilibrium and, on the other hand, has mathematical properties which describe the system's stability.

The broad varieties of population model which abound in the literature are also discussed, with reference to the differences and similarities between models where growth is a continuous process, and those with discrete generations; between models where the population variable is a continuous one, and those where animal numbers come in integral units (demographic stochasticity); between models where the environment is deterministically unvarying, and those where it fluctuates randomly (environmental stochasticity).

9

# WHAT USE ARE GENERAL MODELS?

As has been pointed out by many people, model building in population biology, as in other disciplines, admits a variety of broadly different approaches. Thus Holling (1966, 1968a) has distinguished between relatively detailed "tactical" models and relatively general "strategic" ones, and Levins (1966, 1968b) has indicated a classification in terms of the qualities of "realism, precision and generality." The interest in such questions is shown by the number of people who have taken up Levins' theme and elaborated it, occasionally in discussions which (to paraphrase Voltaire's remark about the Holy Roman Empire) are neither realistic, nor precise, nor general.

The basic fact, perhaps best put by Holling, is that there is a continuous spectrum of possible models, ranging from empirical ones which aim to be of practical use, to rather abstract ones which aim to give qualitative general insights.

At one end of this spectrum are models which strive for a detailed and pragmatic description of quite specific systems. Such "systems analysis" along the lines laid down by Watt (1963, 1966, 1968), Dale (1970), Patten (1971), and Conway (1972), and exemplified by the work of Holling (1965), Conway (1970, 1971), Conway and Murdie (1972), or Hall, Cooper, and Werner (1970), in able hands offers considerable promise both for particular projects of resource management and as a method of codifying otherwise indigestible masses of experimental data. (Be it added that some other massive computer studies could benefit most from the installation of an on-line incinerator.) Conversely, this "tactical" approach does not seem conducive to yielding general ecological principles, nor does it claim to.

At the opposite end of the spectrum is the "strategic" approach, which sacrifices precision in an effort to grasp

10

at general principles. Such general models, even though they do not correspond in detail to any single real community, aim to provide a conceptual framework for the discussion of broad classes of phenomena. Such framework can serve a useful purpose in indicating key areas or relevant questions for the field and laboratory ecologist, or simply in sharpening discussion of contentious issues.

Midway along this continuum of approaches lie such works as Williamson's (1972) book, which covers theoretically based analytic tools for dealing with the population dynamics of specific situations.

Such a continuum of approaches is familiar in other areas of science. For example, the basic pedagogic element of solid state physics is the "perfect crystal." Although nature knows no perfect crystals, the model not only is a useful core for the subject but also provides a springboard for such tactical forays as the optimal choice of superconducting alloys, or impurities in semi-conductors, where the detailed work is often frankly empirical. As in ecology, one must be careful about the circumstances to which the perfect crystal model is applied. While it gives an excellent description of many phenomena, it overestimates the strength of materials by factors of $10^4$; such strength is set by crystal imperfections, and the perfect crystal is therefore ludicrously inadequate here. In like manner, at one end of the spectrum of approaches, the periodic table provides a rough guide to the interaction properties of the chemical elements (and the structure of the periodic table in turn follows from the symmetries of the coulomb force), while, at the other end of the spectrum, industrial chemistry is animated by a more pragmatic approach to the kinetics of molecular reactions. Other such paradigms abound. Sympathetically handled, tactical and strategic approaches mutually reinforce, each providing new insights for the other.

11

In ecology, I think it is true that tactical models of the systems analysis kind, applied to specific individual problems of resource and environmental management, have been more fruitful than has general theory, and they are likely to remain so in the near future. But in the long run, once the "perfect crystals" of ecology are established, it is likely that a future "ecological engineering" will draw upon the entire spectrum of theoretical models, from the very abstract to the very particular, just as the more conventional branches of science and engineering do today.

As may be obvious from these defensive remarks, the mathematical models for biological communities which are treated in this book are all of the very general, "strategic" kind. They are at best caricatures of reality, and thus have both the truth and the falsity of caricatures.

# CHAPTER TWO

# Mathematical Models and Stability

## THE MEANINGS OF STABILITY

A variety of ecologically interesting interpretations can be, and have been, attached to the term "stability."

The most common meaning corresponds to *neighborhood stability,* that is, stability in the vicinity of an equilibrium point in a deterministic system. This circumstance is not only the most tractable mathematically, but also (as we shall see) it often relates to more general stochastic situations, or to large amplitude disturbances.

For population models in deterministic environments, with the environmental parameters all well-defined constants, one is interested in the community equilibria where all the species' populations have time-independent values, that is where all net growth rates are zero. Such an equilibrium may be called stable if, when the populations are perturbed, they in time return to their equilibrium values; the return may be achieved either as damped oscillations or monotonically. Conversely, if such a disturbance tends to amplify itself, the system may be called unstable; again such instability may appear as oscillatory or as monotonic growth in the disturbance. The general cases of stability and instability are divided by the razor's edge of neutral stability, where the perturbed system either remains stationary or oscillates with a constant amplitude set by the magnitude of the initial disturbance. The pathological

13

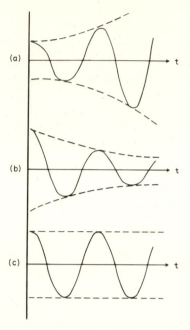

FIGURE 2.1. Schematic illustration of a deterministic, mechanical system which when disturbed from equilibrium is (a) unstable, (b) stable, (c) neutrally stable.

"frictionless pendulum" exhibits neutral stability. These remarks are illustrated by Figure 2.1.

The graphical visualization of these ideas is familiar. The solutions of the equations of population dynamics for a community of $m$ species may be represented as lying on some $m$-dimensional surface. Each point on this landscape corresponds to a set of populations. The equilibrium states are in principle those points where the landscape is flat: hilltops and valley-bottoms. However, the equilibria on hilltops are obviously unstable, unable to survive even the smallest displacement; the valley-bottoms are the stable configurations.

In a linearized or neighborhood stability analysis, one

14

first identifies the equilibrium points, and then looks at the landscape in their immediate neighborhood. Straight-forward mathematical tools, set out below, exist to accomplish the task. In this way equilibrium configurations may be classed as stable or unstable, at least with respect to small amplitude disturbances. This usage of the terms follows that in mechanical systems, and in much of the mathematical genetics of Fisher, Haldane, and Wright.

More generally, since the equations of population biology are nonlinear, the landscape may be quite complicated, and a neighborhood stability analysis may give a misleading representation of the full *global stability* of the system. Thus if our locally stable valley-bottom is, as it were, nestled in the tip of a volcano-like peak, then, although small population perturbations will settle back to the valley floor, a large perturbation may carry the community over the crater's lip, to spill out onto the terrain below. Thus a global stability analysis will seek to comprehend the stability structure of the entire landscape, rather than just the neighborhood of equilibrium points.

If the underlying dynamical equations are linear, which they often are in the physical sciences but essentially never are in population biology, neighborhood and global stability are identical. Moreover, many biologically interesting models, although nonlinear, correspond to relatively simple such landscapes, with one valley (or one hilltop) whose sides slope ever upward (or downward). Obviously in this event the neighborhood stability analysis correctly describes the global stability. Such circumstances are characterized by the existence of a "Lyapunov function," a function $V(N_1, N_2, \ldots, N_m)$ of the population variables with the property that $V$ is positive definite, and $dV/dt$ is negative semi-definite (stable valley) or positive semi-definite (unstable hilltop), throughout the region of population space in question. In other words, the existence of this entity throughout a region $R$ corresponds to the statement that the global stability is legitimately characterized by the neighborhood analysis throughout $R$. (In conventional mechanics, the Hamiltonian constitutes a Lyapunov function. There is, unfortunately, no general way of telling whether a Lyapunov function exists in a given situation, nor of constructing it if it does exist. Substantially more technical reviews of this topic are given lucidly by Burton (1969), Barnett and Storey (1970, Ch. 5), and Rosen (1970, Ch. 3).)

Another complication arising from the nonlinearity of the equations which describe the dynamics of interacting populations is that the equilibrium or steadily maintained system need not necessarily be a point equilibrium (as it must be for a linear system), but can alternatively be a stable limit cycle, wherein the population numbers undergo well-defined cyclic changes in time. For a stable limit cycle, just as for a stable point equilibrium, the system if disturbed will tend to return to the equilibrium configuration. This explicitly nonlinear phenomenon is discussed more fully in Chapter 4. The persistent oscillations of a stable limit cycle are altogether different from, and are not to be confused with, the pathological, sustained oscillations of neutrally stable "frictionless pendulums," such as the classical Lotka-Volterra predator-prey model.

So far in our discussion, stability, whether neighborhood or global, has been a yes-no affair. However a neighborhood stability analysis may distinguish relative degrees of stability by describing whether the locality of the valley-bottom is relatively steeply sloped (making for relatively swift return to equilibrium following a perturbation), or relatively flat. Clearly on a multidimensional surface there will be different gradients in different directions, and the overall stability in the neighborhood of the equilibrium point is likely to be set by the slope of least magnitude.

These "slopes," of course, correspond to the damping rates in an analytic treatment.

The above discussion rests on the assumption that the environmental parameters in our model equations are immutable constants. In reality all such parameters will, to a greater or lesser degree, exhibit random fluctuations. Real environments are uncertain, stochastic. In the deterministic case, we spoke of *the* equilibrium populations, and tested their stability with regard to the imposition of small disturbances. In the stochastic case, there is a continual spectrum of such perturbations and fluctuations built into the fabric of the model, and we speak of equilibrium if there are finite average populations around which the animal numbers fluctuate with steady average variances. The populations are now described in probabilistic terms. A necessary, but insufficient, condition for the persistence of such an equilibrium probability distribution is that the corresponding deterministic population model be stable.

In such stochastic circumstances, it intuitively seems sensible to refer to those systems characterized by large fluctuations in the population numbers as "unstable," and to those with relatively small fluctuations as "stable." As discussed in Chapter 5, this usage is often related in a well-defined way to the relative degree of stability in the deterministic mechanical models.

A yet more general alternative meaning that can be attached to stability in ecological contexts is *structural stability*. This refers to the qualitative effects upon solutions of the model equations produced by gradual variation in the model parameters themselves, that is by modifications in the structure of the basic equations. If the solutions change in a continuous manner (i.e. if the perturbed system is topologically isomorphic to the unperturbed one), the system is said to be structurally stable. Conversely if gradual changes in the system parameters, such as altera-

17

tions in physiographical factors in the biome, as manifested for example by intrusion of glacial tongues, produce qualitatively discontinuous effects, the system is structurally unstable. Most neutrally stable models will be structurally unstable, with slight changes in the basic equations precipitating the system into the category stable or unstable.

One very general approach to the problems of population ecology is to look beyond the stability of the individual populations constituting the system, to see if some general quantities such as net number of species, or overall energy or biomass flow, are roughly conserved. That is, one could look beyond the details of the dynamical stability surface, which may be pocked with valleys and ridges like the surface of the moon, to seek whether there is some broad region of this dynamical space within which the system as a whole may be bounded. This is a wider, if fuzzier, question than is dealt with in this book. The tools of structural stability analysis may well be the appropriate ones to use on this interesting question, which has received little attention.

The recent proof by Smale (1966) that structurally stable systems are, in a precise sense, rare in more than three dimensions could have implications in many biological fields. For a fuller account, albeit mainly in a morphogenetic context, see Thom (1969, 1970). Although we do not go beyond these vague references to structural stability here, the subject is mentioned because it is relevant to the issues under review, and is likely to be one of the growth points of theoretical biology.

For a more thorough review of the meanings that may be given to stability, see Lewontin (1969).

We now turn to focus on one small corner of the large picture sketched above, namely neighborhood stability in deterministic models. The mathematical formalism, presented in the next section, underpins Chapters 3 and 4 explicitly, and remains relevant to the randomly fluctuating environments of Chapters 5 and 6.

## THE COMMUNITY MATRIX

Consider a community comprising but one species, with population $N(t)$, whose dynamics are described by the differential equation

$$\frac{dN(t)}{dt} = F(N(t)). \tag{2.1}$$

Here the population growth rate, $dN/dt$, is given by some function of the populations at time $t$, $F(N)$. Any possible equilibrium population, $N^*$, will by definition be a solution of the algebraic equation obtained by setting the growth rate zero:

$$0 = F(N^*). \tag{2.2}$$

An analysis of small disturbances about the equilibrium population proceeds by first writing the perturbed population as

$$N(t) = N^* + x(t). \tag{2.3}$$

Here $x(t)$ measures the perturbation to the equilibrium population, and is by assumption initially relatively small. An approximate differential equation for this perturbation measure is then obtained by a Taylor expansion of the basic equation (2.1) about the equilibrium point, neglecting terms of order $x^2$ and higher:

$$\frac{dx(t)}{dt} = a\, x(t). \tag{2.4}$$

The quantity $a$ is the derivative,

$$a = (dF/dN)^*, \tag{2.5}$$

19

evaluated at the equilibrium point $N = N^*$. It measures the per capita population growth rate in the immediate neighborhood of the equilibrium point.

The solution of the linearized equation (2.4) is, of course,

$$x(t) = x(0)e^{at}, \tag{2.6}$$

where $x(0)$ is the initial small perturbation. Obviously if $a < 0$, the disturbance dies away exponentially, whereas for $a > 0$ the perturbation grows, and the special case $a = 0$ gives neutral stability. In short, the neighborhood stability analysis gives the equilibrium point at $N^*$ to be stable if and only if $a$ is negative.

As an example, consider the equation of logistic population growth

$$\frac{dN}{dt} = rN(1 - N/K). \tag{2.7}$$

Here the conventional quantity $r$ measures the intrinsic per capita growth rate, and $K$ the total carrying capacity. The possible equilibrium points, from equation (2.2), are $N^* = K$ and $N^* = 0$. For the point at $N^* = K$, we find $a = -r$, corresponding to stability if and only if $r > 0$; conversely the neighborhood of the point $N^* = 0$ is stable if and only if $r < 0$. For this simple example a full nonlinear solution of equation (2.7) is easy and familiar, and we remark that for $r > 0$ the neighborhood analysis gives a true description of the global stability throughout the relevant domain $N \geqslant 0$, namely a stable equilibrium population of magnitude $K$. (A Lyapunov function, $V(N) = (1 - N/K)^2$, can be written down here, and therefore the neighborhood stability analysis describes the global stability.)

The case of a multispecies community, with $m$ populations $N_i(t)$ labeled by indices $i = 1, 2, \ldots, m$, is in principle equally straightforward.

The essential thing which emerges is the community matrix, $A$. Just as in the single species system the "1 × 1 matrix" $a$ both summarizes the biology (being the per capita growth rate near equilibrium) and sets the neighborhood stability (by its sign), so too in the multispecies system the $m \times m$ matrix $A$ both summarizes the biology (its elements being determined by the interactions between and within species near equilibrium) and sets the neighborhood stability (by the sign of its eigenvalues). An awareness of these general features of the community matrix is really all that is required to follow the basic themes in the subsequent chapters, which may comfort anyone for whom the mathematics below is not easy.

Suppose the multispecies population dynamics are given by a set of $m$ equations

$$\frac{dN_i(t)}{dt} = F_i(N_1(t), N_2(t), \ldots, N_m(t)). \qquad (2.8)$$

Here the growth rate of the $i$th species at time $t$ is given by some nonlinear function $F_i$ of all the relevant interacting populations. Again the equilibrium populations, $N_i^*$, follow from the $m$ algebraic equations obtained by putting all growth rates zero:

$$0 = F_i(N_1^*, N_2^*, \ldots, N_m^*). \qquad (2.9)$$

Expanding about this equilibrium, for each population we write

$$N_i(t) = N_i^* + x_i(t), \qquad (2.10)$$

where $x_i$ measures the initially small perturbation to the $i$th population. Taylor expanding each of the basic equations (2.8) around this equilibrium, and discarding all terms which are of second or higher order in the population perturbations $x$, a linearized approximation is obtained:

$$\frac{d x_i(t)}{dt} = \sum_{j=1}^{m} a_{ij} x_j(t). \qquad (2.11)$$

2 1

This set of $m$ equations describes the population dynamics in the neighborhood of the equilibrium point. Equivalently, we may write, in matrix notation,

$$\frac{d\ \mathbf{x}(t)}{dt} = A\ \mathbf{x}(t). \tag{2.12}$$

Here $\mathbf{x}$ is the $m \times 1$ column matrix of the $x_i$, and $A$ is the $m \times m$ "community matrix" (Levins 1968a), whose elements $a_{ij}$ describe the effect of species $j$ upon species $i$ near equilibrium. The elements $a_{ij}$ depend both on the details of the original equations (2.8), and on the values of the equilibrium populations, according to the recipe

$$a_{ij} = \left(\frac{\partial F_i}{\partial N_j}\right)^*. \tag{2.13}$$

The partial derivatives, denoting the derivatives of $F_i$ keeping all populations except $N_j$ constant, are to be evaluated with all populations having their equilibrium values.

For the set of linear differential equations (2.11) the solution may be written

$$x_i(t) = \sum_{j=1}^{m} C_{ij} \exp(\lambda_j t). \tag{2.14}$$

This is the multispecies analogue of the single species solution (2.6). The $C_{ij}$ are constants which depend on the initial values of the perturbations to the populations, and the time dependence is contained solely in the $m$ exponential factors. The $m$ constants $\lambda_j$ (with $j = 1, 2, \ldots, m$), which obviously characterize the temporal behavior of the system, are the so-called eigenvalues of the matrix $A$. They are found by substituting (2.14) into (2.11) to get

$$\lambda\ x_i(t) = \sum_{j=1}^{m} a_{ij} x_j(t), \tag{2.15}$$

or, in the more compact matrix form,

22

$$(A - \lambda I)\, \mathbf{x}(t) = 0. \tag{2.16}$$

Here $I$ is the $m \times m$ unit matrix. This set of equations possesses a non-trivial solution if and only if the determinant vanishes:

$$\det |A - \lambda I| = 0. \tag{2.17}$$

This is in effect a $m$th order polynomial equation in $\lambda$, and it determines the eigenvalues $\lambda$ of the matrix $A$. They may in general be complex numbers, $\lambda = \zeta + i\xi$; in any one term in equation (2.14) the real part $\zeta$ produces exponential growth or decay, and the imaginary part $\xi$ produces sinusoidal oscillation. Figure 2.1 corresponds to one such term with $\xi \neq 0$ and (a) $\zeta > 0$, (b) $\zeta < 0$, (c) $\zeta = 0$. Looking back at equation (2.14), it is clear that the perturbations to the equilibrium populations will die away in time if, and only if, all eigenvalues $\lambda$ have negative real parts. If any one eigenvalue has a positive real part, that exponential factor will grow ever larger as time goes on, and consequently the equilibrium is unstable. The special case of neutral stability is attained if one or more eigenvalues are purely imaginary numbers, and the rest have negative real parts.

Collecting these remarks, we observe that an equilibrium configuration in the multispecies system will have neighborhood stability if, and only if, all eigenvalues of the community matrix lie in the left-hand half of the plane of complex numbers. This criterion is illustrated in Figure 2.2. As a final notational flourish, it is convenient to define $\Lambda$ as minus the largest real part of all the eigenvalues of the community matrix:

$$-\Lambda = [\text{Real } (\lambda)]_{\text{max}}. \tag{2.18}$$

The stability criterion then becomes

$$\Lambda > 0. \tag{2.19}$$

Fulfillment of the neighborhood stability conditions of

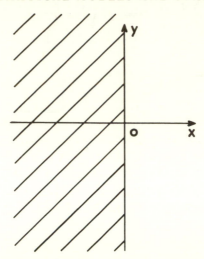

FIGURE 2.2. The eigenvalues $\lambda = x + iy$ of the community matrix $A$ may be represented as points $(x, y)$ in the complex plane. In a deterministic environment with population growth a continuous process, the criterion for an equilibrium community to be stable with respect to small disturbances is that all such eigenvalues have negative real parts, that is lie in the hatched region.

Figure 2.2 corresponds graphically to the dynamical landscape sloping in every direction upward from the equilibrium point; the magnitude in the direction of least slope is measured by the real part of the eigenvalue nearest to the imaginary axis in Figure 2.2, that is by $\Lambda$.

If one or more eigenvalues have positive real parts (i.e., if $\Lambda < 0$), all we can say with certainty is that there is not a stable equilibrium point. Perturbations will initially grow, but the neighborhood analysis leaves their ultimate fate uncertain. Eventually terms of order $x^2$ and higher become important, and nonlinearities decide whether the perturbations will grow until extinctions are produced, or whether the system may settle into some limit cycle. Likewise even if the equilibrium point is stable to small perturbations, as shown by the neighborhood analysis, its response to severe buffetings is not necessarily known.

Appendices I and II may be referred to at this point. Appendix I is for the benefit of those who may find the above presentation rather abstract, and it contains explicit examples to illustrate the analysis in relatively familiar circumstances. Conversely, Appendix II goes beyond the above discussion, both to give some general rules relating to whether the eigenvalues of the community matrix all lie in the left half of the complex plane, and to catalogue the eigenvalues of some ecologically interesting matrices. We shall frequently refer to Appendix II throughout the succeeding chapters.

The community matrix $A$ is clearly the central figure in this section. It is a quantity of direct biological significance. Its elements $a_{ij}$ describe the net effect of species $j$ upon species $i$ near equilibrium. A diagram of the trophic web immediately shows which elements $a_{ij}$ are zero (no web link); the type of interaction sets the sign of the non-zero elements; and the details of the interactions determine the magnitude of these elements. The sign structure of this $m \times m$ matrix is directly tied to Odum's (1953) scheme, which classifies interactions between species in terms of the signs of the effects produced. He characterizes the effect of species $j$ upon species $i$ as positive, neutral, or negative (that is, $a_{ij}$ +, 0, or −) depending on whether the population of species $i$ is increased, is unaffected, or is decreased by the presence of species $j$. Thus for the pair of matrix elements $a_{ij}$ and $a_{ji}$ we can construct a table of all possible interaction types:

|  | | Effect of species $j$ on $i$ [i.e. sign of $a_{ij}$] | | |
|---|---|---|---|---|
|  | | + | 0 | − |
| Effect of species | + | ++ | +0 | +− |
| $i$ on $j$ | 0 | 0+ | 00 | 0− |
| [i.e. sign of $a_{ji}$] | − | −+ | −0 | −− |

Apart from complete independence, there are five distinguishably different categories of interaction between any given pair of species, namely commensalism (+0), amensalism (−0), mutualism or symbiosis (++), competition (−−), and general predator-prey (+−) including plant-herbivore, parasite-host, and so on. For a more thorough exposition, see Williamson (1972, Ch. 9).

In conclusion, it may well be remarked that for a system with two species it is often possible to elucidate the full nonlinear topology of the "phase space" in which the point representing the two populations moves, and thereby to effect a global stability analysis. Most ecology books present such a phase plane analysis for the Lotka-Volterra equations for two competitors, and Slobodkin (1961, Figure 7-2) gives neat sketches of the actual stability landscapes. This geometrical technique becomes less useful as one moves beyond two species and two dimensions, partly for human reasons (book pages are two dimensional), and partly for mathematical reasons (the topology of two-dimensional surfaces can be different in qualitative ways from that of higher dimensions, a point discussed in Chapter 4 in connection with the Poincaré-Bendixson theorem). Therefore, as our main interest is in multispecies communities, we have concentrated on presenting analytic techniques of neighborhood stability analysis.

## VARIETIES OF POPULATION MODELS

In modeling biological populations, various approaches differing in matters of biological and technical detail can be distinguished.

### Discrete versus Continuous Growth

One interesting distinction is between models where population growth is a continuous process (as for example in human populations), and those with discrete generations or age classes (as for example among salmon or cicadas).

So far we have dealt with deterministic models in which population growth is a continuous process. That is, the independent variable $t$ is continuous, and we have differential equations. A paradigm for this circumstance is the simple single species exponential growth equation,

MODEL I:
$$\frac{dN(t)}{dt} = r N(t).$$
(2.20)

This has the familiar solution

$$N(t) = N_o e^{rt},$$
(2.21)

with $N_o$ the initial population at $t = 0$.

If there are separate generations, then the independent variable $t$ is a discrete one, and we have difference equations for the discretized growth rates $N(t + \tau) - N(t)$. The paradigm corresponding to Model I will have the form

MODEL II:
$$N(t + \tau) = (1 + r\tau) N(t),$$
(2.22)

where $\tau$ is the time interval between successive generations. The time taken for $k$ generations to elapse is thus $t = k\tau$, and the population then is

$$N(t = k\tau) = N_0(1 + r\tau)^k.$$
(2.23)

In the limit as the generation time tends to zero, that is as it becomes much smaller than all other relevant times in the system, we have

$$\lim_{\tau \to 0}(1 + r\tau)^{t/\tau} \to e^{rt}.$$
(2.24)

Thus, as the time interval between successive generations becomes negligible, we recover the continuous growth result (2.21), as we obviously should.

Similarly, in the general multispecies situation, a relation may be established between the stability properties of population models with continuous, and with discrete, growth (May, 1972a).

Corresponding to any particular differential equation model is an analogous ("homologous") difference equation model, in which all the biological features such as trophic structure, birth and death rates, competitive and prey-predator interactions, and so on are identical, save only that population growth takes place at discrete intervals, rather than as a continuous process. For the multispecies community whose dynamics are described by the set of $m$ differential equations (2.8), the homologous difference equations are

$$N_i(t + \tau) - N_i(t) = \tau \, F_i(N_1(t), N_2(t), \ldots, N_m(t)), \quad (2.25)$$

with $\tau$ the generation time. Again the possible equilibrium populations, $N_i^*$, are determined from equation (2.9). This identity between the models in the static, time-independent limit follows from our definitions: if time does not enter into consideration, the difference between continuous and discrete growth processes is irrelevant. The stability of this equilibrium with respect to small disturbances may be studied by methods analogous to those set out above for systems with continuous growth. This is done in Appendix III.

The upshot of this neighborhood analysis is that stability hinges on the eigenvalues of identically the *same* community matrix as for the homologous system of differential equations (2.8). But, whereas the continuous growth system was stable if and only if all these eigenvalues lay in the left half complex plane, stability of the corresponding system with discrete generations requires the more stringent condition that all these same eigenvalues lie inside a circle of radius $1/\tau$, centered at $-(1/\tau)$ on the real axis, in the complex plane. This more severe criterion is illus-

28

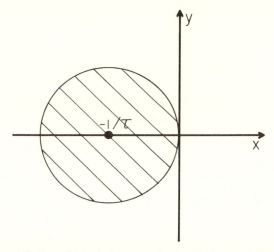

FIGURE 2.3. For a biological community in which growth is a discrete process, as represented by the system of difference equations (2.25), the criterion for stability in the neighborhood of an equilibrium point is that all eigenvalues of the *same* community matrix as in Figure 2.2 (i.e. equation (2.13)) lie inside the hatched circle in the left half of the complex plane.

trated in Figure 2.3, which is to be contrasted with Figure 2.2.

It is widely understood that difference equations tend to be less stable than their differential equation twins, because the finite time lapse between generations of growth will have the destabilizing effects associated with any time lag in an interactive system (see, for example, Bartlett, 1960, Chs. 4–6, or Maynard Smith, 1968, Ch. 2). The comparison between Figures 2.2 and 2.3 makes this quite explicit; clearly stability of the difference equation system implies stability of the differential equation one, but the converse is not necessarily true.

Thus, by way of a simple example, the discrete generations version of the logistic growth equation (2.7) is

$$N(t + \tau) - N(t) = \tau r N(t)[1 - N(t)/K]. \qquad (2.26)$$

For the possible equilibrium population value $N^* = K$, the $1 \times 1$ community matrix was earlier seen to have the eigenvalue $-r$. From Figure 2.3, the neighborhood stability criterion for this equilibrium population at $K$ is now $2/\tau > r > 0$. This contrasts with the requirement $r > 0$ for the model with continuous growth. Clearly as $\tau$ tends to zero the two models coincide, but, for any finite generation time $\tau$, too large a per capita intrinsic growth rate $r$ can lead to diverging oscillations in models with discrete growth intervals.

It is also clear how Figure 2.2 is attained as a limiting case of Figure 2.3. As the interval $\tau$ between successive population growth steps becomes smaller and smaller compared to other relevant time scales in the system, the radius $(1/\tau)$ of the circle becomes larger and larger, and the center (at $-1/\tau$) recedes ever further to the left, until the right-hand boundary of the circle is for most practical purposes coincident with the imaginary axis, leading to Figure 2.2.

Without this synthesis, one is liable to end up comprehending the relation between continuous and discrete growth in individual models, one by one, in a rather unsatisfactory way.

In any particular application, the choice between continuous time models and discrete ones should, of course, be dictated by the biological realities. Although the bulk of this book is confined to models with continuous growth, and, consequently, differential equations, analogous results could be presented for difference equation systems. One could, as it were, write a homologous book.

### Demographic Stochasticity

Another interesting distinction is between a description where the dependent variable, the total population $N(t)$, changes continuously, and one in which it changes

30

in integral steps. In the former case we speak of the number of animals present at time $t$, $N(t)$, and in an infinitesimal time $dt$ this quantity will change by an infinitesimal amount $dN$; the models are deterministic. In the latter case, animals come only in integer units, and we have a distribution function, $f(n, t)$, which gives the probability to find $n = 0$, 1, 2, ..., $N$, ... animals at time $t$; the models are stochastic. By taking the usual statistical moments of the probability distribution, we get the mean number of animals at $t$ (which may, or may not, be identical with the deterministic $N(t)$), the variance in the number, and so forth.

The stochastic paradigm corresponding to Model I, where growth is a continuous process, is given by the following probabilities for an individual to give birth to offspring in an infinitesimal time interval $dt$:

$$\text{MODEL III:} \quad \left. \begin{array}{l} \text{probability 1 offspring} = r\,dt \\[2mm] \text{probability 0 offspring} = 1 - r\,dt \end{array} \right\} \quad (2.27)$$

If the initial population contains $N_o$ individuals, the probabilities to find $n$ individuals at time $t$ ($n = N_o$, $N_o + 1$, ...) are given by the distribution function (e.g. Bartlett 1966, p. 74)

$$f(n, t) = \frac{(n-1)!}{(N_o - 1)!(n - N_o)!} \, e^{-N_o rt}[1 - e^{-rt}]^{n-N_o}. \quad (2.28)$$

This result is, for Model III, the analogue of equation (2.21). Thence we may calculate the average population,

$$\langle n(t) \rangle = N_o e^{rt}, \quad (2.29)$$

the variance,

$$\langle (n - \langle n \rangle)^2 \rangle = N_o e^{2rt}(1 - e^{-rt}), \quad (2.30)$$

and so on.

Clearly the stochastic approach gives a fuller description of the system, bought at the expense of harder calculation. The essential difference is that in the deterministic model each member of the population gives birth to some tiny fraction of an individual (!) in each small interval of time, whereas in the stochastic model only whole animals are born, with specified probabilities. For very large populations this distinction becomes unimportant. In the canonical models above we see that the stochastic mean (2.29) is identical with the deterministic population (2.21), and the statistical root-mean-square relative fluctuation about this mean is

$$\frac{\{\langle (n - \langle n \rangle)^2 \rangle \}^{1/2}}{\langle n \rangle} = N_o^{-1/2}(1 - e^{-rt})^{1/2}. \qquad (2.31)$$

That is,

$$\lim_{t \to \infty} \frac{\sqrt{\text{variance}}}{\text{mean}} \to N_o^{-1/2}. \qquad (2.32)$$

Thus for very large populations, $N_o \gg 1$, the deterministic approach should serve.

Alternatively, if growth is a discrete process, the stochastic analogue of the paradigm II has birth occurring at specific times, at intervals $\tau$ apart, with probabilities

MODEL IV:
$$\left.\begin{array}{l} \text{probability 1 offspring} = r\tau \\[2mm] \text{probability 0 offspring} = 1 - r\tau \end{array}\right\} \qquad (2.33)$$

The resulting probability distribution at time $t = k\tau$, after $k$ generations, has the first moment (the mean)

$$\langle n(t = k\tau) \rangle = N_0(1 + r\tau)^k. \qquad (2.34)$$

The second moment (the variance) at $t = k\tau$ is

$$\langle (n - \langle n \rangle)^2 \rangle = N_0(1 - r\tau)(1 + r\tau)^{2k-1}[1 - (1 + r\tau)^{-k}], \qquad (2.35)$$

and so on.

As in the comparison between Models I and II, these equations reduce to those of Model III as $\tau$ tends to zero (see equation (2.24)). As in the comparison between I and III, the stochastic average (2.34) is identical with the deterministic population value (2.23), and again the root-mean-square relative fluctuations are negligible when the population is large:

$$\lim_{t\to\infty} \frac{\sqrt{\text{variance}}}{\text{mean}} \to N_o^{-1/2} \left(\frac{1-r\tau}{1+r\tau}\right)^{1/2}. \qquad (2.36)$$

Our standard example, the logistic growth formula (2.7), may also be discussed in such a stochastic framework (Bartlett, 1960, Ch. 4). Again the conclusion is that, as long as the carrying capacity $K$ corresponds to a large number of animals, the population fluctuations induced by a statistical treatment are characterized by a relative amplitude proportional to $K^{-1/2}$.

The generalized central limit theorems of McNeil and Schach (1971) further substantiate the fact that for large populations, $N \gg 1$, the averages tend to be proportional to $N$, and the variances to $N$, so that the root-mean-square relative fluctuations scale as $N^{-1/2}$.

Thus so long as *all* relevant populations in the food web are reasonably large, the deterministic approach typified by Models I and II should suffice.

A more detailed justification for using deterministic, rather than stochastic, equations in dealing with large populations in an ecological context has been given by Beverton and Holt (1956) and Watt (1968, p. 350). The cautionary notes sounded, for example, by Becker (1973) are mainly in circumstances where at least one population is small.

These stochastic features, arising because the population variable is fundamentally a discrete one, have been christened "demographic stochasticity."

The variety of models typified by the above paradigms may be summarized schematically:

Dependent Variable, $N$

|  |  | continuous | discrete |
|---|---|---|---|
| Independent | continuous | Model I | Model III |
| Variable, $t$ | discrete | Model II | Model IV |

In this table, one passes from right to left as the populations become very large ($N \gg 1$), and from bottom to top as the growth steps become small ($\tau \to 0$). The subsequent chapters mainly treat the kind of model typified by I.

### Environmental Stochasticity

So far, the environmental parameters (typified by the growth rate $r$ in our paradigms I–IV) have been taken to be constant, unvarying in time. More realistically, such environmental parameters should be time-dependent.

An example would be the modification of the paradigm I to read

MODEL V:
$$\frac{dN(t)}{dt} = r(t)\, N(t). \qquad (2.37)$$

If $r(t)$ is some prescribed function of $t$, no essential new feature enters. But if $r(t)$ is fluctuating randomly about some mean value, as it usually is in nature, substantial complications enter. A good review is due to Sykes (1969).

The consequent element of environmental stochasticity generates population fluctuations whose relative magnitude is set by the degree of environmental variance, independent of the absolute population size. This is to be contrasted with the effects of demographic stochasticity, where

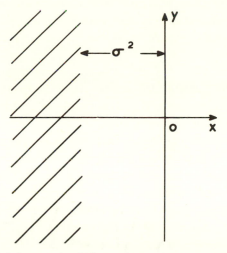

FIGURE 2.4. In a randomly fluctuating environment, with variance characterized by $\sigma^2$, a rough criterion for the population numbers not to fluctuate to extinction is that all eigenvalues of the (average value) community matrix of equation (2.13) lie in the hatched region, a distance of the order of $\sigma^2$ to the left of the imaginary axis.

the relative magnitude of the fluctuations characteristically scales as $N^{-1/2}$.

This question is taken up in Chapter 5. The community matrix will be seen often to retain its relevance. For the population fluctuations to be not too severe, so that the community persists, a rough criterion is that all the eigenvalues of $A$ lie not merely in the left half complex plane, but to the left of the imaginary axis by an amount $\sigma^2$ which characterizes the environmental variance. This qualitative criterion is illustrated in Figure 2.4.

## SUMMARY

The variety of meanings which may sensibly be attached to stability in ecological contexts are reviewed. Neighborhood stability is fastened upon as being mathematically the most tractable, and often underlying other more complicated usages. Explicitly nonlinear phenomena, such as limit cycles, are raised with a view to later discussion.

Within the formal framework of neighborhood stability analysis, the community matrix is held up as an entity which both epitomizes the biology of the community and also sets its stability character.

A diversity of population models are put in perspective. We choose henceforth to use models where growth is continuous (leading to differential rather than difference equations), and where the populations are numerically large (thus avoiding complications of demographic stochasticity). The effects of environmental stochasticity, which are independent of population size, will be included.

36

# CHAPTER THREE

# Stability versus Complexity in Multispecies Models

One of the central themes of population ecology is that increased trophic web complexity leads to increased community stability.

A good deal of evidence has been assembled (e.g. Elton, 1958, pp. 145–150; Pimentel, 1961) to show that, in nature, species population stability is typically greater in structurally complex communities than in simple ones. Elton's review points out that both mathematical models and laboratory experiments on simple one-predator–one-prey systems oscillate violently; that cultivated or planted land, or orchards, or the simpler communities on islands have shown themselves to be comparatively unstable; whereas the rain forest, the paradigm of trophic web complexity, appears very stable. Hutchinson (1959), referring to Odum (1953) and MacArthur (1955), notes that "oscillations observed in arctic and boreal fauna may be due in part to the communities not being sufficiently complex to damp out oscillations," a point of view which has been elaborated by subsequent authors (e.g., Macfadyen, 1963, p. 182).

The hypothesis that increased food web complexity causes increased stability has, on occasion, been accorded the status of a mathematical theorem. MacArthur's (1955) suggestion that community stability may be roughly proportional to the logarithm of the number of links in the trophic web has sometimes been wishfully mistaken for such a theorem. In fact, this suggestion rests on the valid argument, borrowed from information theory, that such a logarithm measures the degree of organization (complexity) of the web, followed by the usual intuitive association between complexity and stability. This work, elegant and insightful though it is, is *not* a "formal [mathematical] proof of the increase in stability of a community as the number of links in its food web increases" (Hutchinson, 1959, p. 103).

With the contemporary upsurge of interest in these questions, accumulating evidence suggests that the relation between complexity and stability is substantially more complicated than appears at first sight.

In arguing for a reappraisal of the question, Watt (1968, p. 43) has observed that "it is a disturbing fact that many of the most historically important pest species (rodents, locusts, grasshoppers, and forest-insect defoliators such as the spruce budworm) are attacked by an enormous number and variety of species. Grasshoppers, for example, are attacked by a variety of pathogens, mites, nematodes, spiders, wasps, tachinid and sarcophagid fly parasites, egg parasites, praying mantises, snakes, mammals, and birds — it is truly a wonder that any grasshoppers ever survive." In discussing pest control strategies, Turnbull and Chant (1961) and DeBach et al. (1964) have argued that a relatively simple predator-prey system can often be more stable than a complex one, and Zwolfer's (1963) study of six species of *Lepidoptera* plus their parasite complexes, although qualitative rather than quantitative, has suggested that in this case the simpler systems are more stable. Hairston et al.'s (1968) investigation of an artificial laboratory system comprising at maximum three bacteria species, three paramecium species, and two protozoan predators led them to "conclude that much more experimental and observational work is necessary before the nature of any functional relationship between diversity and stability can be claimed with confidence." In a cogent paper, South-wood and Way (1970) have stressed the need to consider the detailed character of the particular trophic web structure, before attempting generalizations as to population stability (see their Figure 3 and accompanying discussion).

Elton's argument that isolated oceanic islands are readily invadable is well documented (e.g. Holdgate and Wace, 1971). But, as admitted by Elton (1966), the point really concerns vulnerability, not stability. Nor is such vulnerability confined to the simple ecosystems of islands. The stability of complex continental ecosystems was no armor against the Japanese beetle, European gypsy moth, or Oriental chestnut blight *Endothia parasitica* in North America, to cite a few among many examples. It is trivial, but not irrelevant, to observe that stability was hardly enhanced by the extra links added to the trophic web in these instances. Likewise, removal of one species can lead to a severe collapse in the overall trophic structure: thus Paine (1966) has shown that removal of one species from an intertidal community of marine invertebrates led to its collapse from a 15-species to an 8-species system in under two years. These last two sentences add up to the remark that species integration (Emerson, 1949) is a very nonlinear affair, and complex communities contain much more "information" than can be estimated by counting links in the trophic web. Even the complex and diverse coral reef, commonly thought of as the aquatic analogue of the rain forest, has recently had its stability called into question by *Acanthaster planci* in the Pacific.

In this chapter we examine one tiny piece of this large jigsaw puzzle. It has to do with the relation between simple mathematical models for communities with many, as contrasted with few, species.

The relation between the present mathematical models and the complications of the natural world should be emphasized. In the real world there is, on the one hand, the complicated character of individual interactions between and within species: predator switching, spatial heterogeneity and boundary effects, density-dependent birth and death rates, to name a few. These realistic complications are present even in one- or two-species systems, and can easily stabilize them. On the other hand, there are the complexities consequent upon the inclusion of large numbers of species in our model community. In the present chapter, we restrict attention to the simplest models for individual interactions between species, however unrealistic they may or may not be, and proceed to understand the effects introduced by adding more and more species to the total system. In this way we may hope to get a feeling for the effects of diversity (in the sense of a large number of species) per se.

## SOME GENERAL PREDATOR-PREY MODELS

Elton's (1958) first argument for the complexity-stability thesis is that simple mathematical models of one-predator–one-prey systems do not possess a stable equilibrium, but exhibit oscillatory behavior. This argument is only germane if the analogous mathematical models of many-predator–many-prey systems are correspondingly more stable. The first thing we do here is to investigate such $n$-predator–$n$-prey systems ($n$ large), and we find them to be in general less stable, and never more stable, than the simple two-species model invoked by Elton. This would seem to invalidate the first of Elton's six classic arguments; the other five are not affected. Moreover, the model provides a specific counterexample to any universal use of trophic link counting as a measure of stability. The $n$-predator–$n$-prey system, with $n^2$ links in its web, is at best no more stable than the analogous singly linked one-predator–one-prey system.

The classic model for a deterministic one-predator–one-prey system with continuous growth is that of Lotka (1925) and Volterra (1926)

$$\frac{dH(t)}{dt} = H(t)[a - \alpha P(t)] \qquad (3.1\text{a})$$

$$\frac{dP(t)}{dt} = P(t)[-b + \beta H(t)]. \qquad (3.1\text{b})$$

Here $H(t)$, $P(t)$ are the populations of prey and predator, respectively, at time $t$. The parameter $a$ relates to the birth rate of the prey, $b$ to the death rate of the predator, and $\alpha$, $\beta$ to the interaction between the species: all are positive numbers. These equations constitute one particular model for the single-link trophic web illustrated in Figure 3.1(a). They constitute the simplest representation of the essentials of the nonlinear predator-prey interaction.

Although they may rightly be criticized for lack of detailed realism, it is often not appreciated that equations of this Lotka-Volterra type faithfully characterize the stability properties of a much wider class of models. The choice of the per capita growth rates (the factors in square

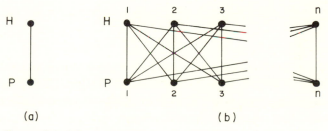

FIGURE 3.1. Schematic representation of a two-level trophic web with (a) one species at each level, and (b) $n$ species at each level. $H$ and $P$ stand for host and parasite, or alternatively for herbivore and predator.

brackets in equations (3.1)) as linear functions of the population variables corresponds to the first approximation in a Taylor series expansion about the equilibrium points in broad classes of more general models, a point emphasized by Lotka (1925, p. 62; 1932), Volterra (1931; 1937) and MacArthur (1970). It is largely for this reason that the valuable ecological "competitive exclusion principle," which originally was derived from the analogous two-competitors Lotka-Volterra equation, is so widely applicable. The feature that *is* objectionable about the equations (3.1) is the absence of intraspecific interaction terms, that is the absence of $H(t)$ inside the square brackets in (3.1a), or of $P(t)$ in (3.1b); the analogous competition equations are free from this particular objection.

The stability character of the equations (3.1) is discussed in Appendix I, which uses this familiar example to illustrate the role played by the community matrix. The prospective equilibrium point clearly has populations $H^* = b/\beta$, $P^* = a/\alpha$. The $2 \times 2$ community matrix may be seen to be

$$A = \begin{pmatrix} 0 & -\alpha b/\beta \\ \beta a/\alpha & 0 \end{pmatrix}. \tag{3.2}$$

The eigenvalues of the matrix are the complex conjugate pair of purely imaginary numbers,

$$\lambda = \pm i(ab)^{1/2}. \tag{3.3}$$

That is, both eigenvalues have real parts equal to zero, corresponding to neutral stability. If displaced from this equilibrium, the system is neither stable (does not tend to return) nor unstable (the disturbance does not grow and grow), but rather it endlessly oscillates or "hunts" around the equilibrium population, as is illustrated in Figure 3.2. The oscillatory period is $2\pi(ab)^{-1/2}$.

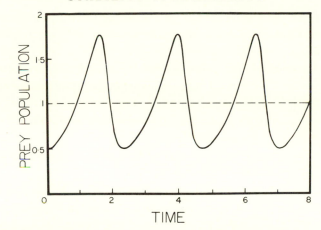

FIGURE 3.2. Oscillations in the population of the prey species according to the simple Lotka-Volterra equation (3.1). The population has been renormalized so that the equilibrium value $H^*$ (denoted by the dashed line) is defined to be unity. The initial conditions are then $H(0) = 0.5$, $P(0) = 1.0$. The coefficients here have the proportions $a = 1$, $b = 10$, $\alpha = \beta = 1$; the absolute magnitude of both the populations, and the time, can of course be rescaled.

This well-known result is the basis for the logically incomplete argument referred to in the opening paragraph of this section.

We now extend this analysis to the case of a system with $2n$ species, namely $n$ predators with the population of the $i$th species being $P_i(t)$ [$i = 1, 2, \ldots, n$], and $n$ prey species with populations $H_i(t)$. The direct analogue of the Lotka-Volterra equation for this $n$-predator–$n$-prey system is

$$\frac{dH_i(t)}{dt} = H_i(t)[a_i - \sum_{j=1}^{n} \alpha_{ij} P_j(t)] \qquad (3.4a)$$

$$\frac{dP_i(t)}{dt} = P_i(t)[-b_i + \sum_{j=1}^{n} \beta_{ij} H_j(t)], \qquad (3.4b)$$

with $i = 1, 2, \ldots, n$. Again, all the parameters $a_i, b_i, \alpha_{ij}, \beta_{ij}$ are positive numbers. For the trophic web with $n^2$ links

43

illustrated in Figure 3.1(b), this set of equations provides a model which is the analogue of equation (3.1) for Figure 3.1(a).

The equilibrium populations $P_i^*$ for the $n$ predator species are given by setting the terms in the square brackets in equation (3.4a) equal to zero. The consequent set of $n$ linear equations may be written compactly in matrix form as

$$\boldsymbol{\alpha} \, \mathbf{P}^* = \mathbf{a}. \tag{3.5a}$$

Here $\boldsymbol{\alpha}$ is the $n \times n$ matrix with elements $\alpha_{ij}$, $\mathbf{P}^*$ the $n \times 1$ column matrix of the equilibrium $P_i^*$, and $\mathbf{a}$ the column matrix of the $a_i$. Similarly for the equilibrium prey populations $H_i^*$,

$$\boldsymbol{\beta} \, \mathbf{H}^* = \mathbf{b}. \tag{3.5b}$$

In order that the model make sense, the coefficients are assumed to be such that equations (3.5) give finite positive values for all populations at equilibrium: this is the only restriction.

The community matrix which characterizes the stability of this multispecies system is now obtained by applying the recipe (2.13) to the equations (3.4). It is evidently a $2n \times 2n$ matrix, partitioned into four $n \times n$ blocks:

$$A = \left( \begin{array}{c|c} \mathbf{0} & -\boldsymbol{\alpha}^* \\ \hline \boldsymbol{\beta}^* & \mathbf{0} \end{array} \right). \tag{3.6}$$

The two diagonal blocks, corresponding to prey-prey and predator-predator interactions, are $n \times n$ null matrices; $\boldsymbol{\alpha}^*$ and $\boldsymbol{\beta}^*$ are $n \times n$ matrices with elements

$$\alpha_{ij}^* = H_i^* \, \alpha_{ij}, \quad \beta_{ij}^* = P_i^* \, \beta_{ij}, \tag{3.7}$$

respectively. This multispecies community matrix for the web of Figure 3.1(b) is to be compared with that of equation (3.2) for Figure 3.1(a). It may be seen that the $2n$ eigenvalues of this matrix occur in $n$ pairs, each pair having the form $\lambda = \zeta + i\xi$, $-\zeta - i\xi$. That is, for every eigenvalue having a negative real part, there is a companion having a positive real part (May, 1971; alternatively the more general result of Appendix II, p. 195, applies here). Consequently, *either* all the eigenvalues have real parts zero, in which case we again have neutral stability, *or* at least one eigenvalue has a positive real part, and the system is unstable.

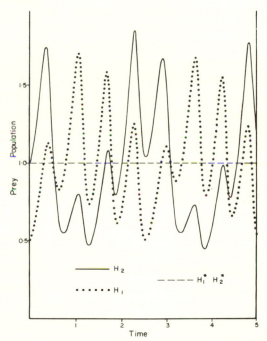

FIGURE 3.3. Oscillations in the prey species populations, $H_1$ and $H_2$, governed by a particular equation of the form (3.4). The populations have been rescaled so that the equilibrium values (denoted by the broken line) are unity. The initial conditions are $H_1(0) = 0.5$, $H_2(0) = P_1(0) = P_2(0) = 1.0$. The coefficients in this particular example have the proportions $a_1 = a_2 = 3$, $\alpha_{11} = \alpha_{22} = 2$, $\alpha_{12} = \alpha_{21} = 1$, $b_1 = 40$, $b_2 = 20$, $\beta_{11} = 3$, $\beta_{12} = \beta_{21} = \beta_{22} = 1$.

45

Thus this $n$-predator–$n$-prey system at best has the same stability properties as the corresponding one-predator–one-prey model, and in general the multispecies system is unstable rather than merely oscillatory.

Figure 3.3 illustrates the behavior of the prey populations in a numerical solution of a 2-predator–2-prey system for the neutrally stable case, that is when all 4 eigenvalues of the community matrix lie on the imaginary axis. The predator populations behave similarly. The initial disturbance from equilibrium in one of the prey species here is the same as that for the (only) prey species in Figure 3.2; the other species are initially unperturbed. We observe that the populations oscillate with amplitudes much the same as for the simple one-predator–one-prey system illustrated in Figure 3.2. Since two frequency components are present in Figure 3.3, in contrast with the single frequency (3.3) in Figure 3.2, the oscillatory pattern is more complicated. More generally, a purely oscillatory $n$-predator–$n$-prey system will have an oscillatory pattern synthesized from $n$ frequency components, and the amplitude of oscillation will still usually be similar to that for a one-predator–one-prey system with similar initial conditions.

Figure 3.4 illustrates the prey populations in a numerical solution for the case of an unstable 2-predator–2-prey system, that is when at least one eigenvalue of the community matrix lies in the right half of the complex plane (cf. Figure 2.2). The behavior of the two predator populations is similar, with one first being eliminated, and the other then settling down to oscillate. Here the equilibrium conditions, obtained by solving equations (3.5), have little relevance, as any small disturbance will grow in time until species are eliminated and the system simplifies itself to a one-predator–one-prey community, which is then necessarily purely oscillatory.

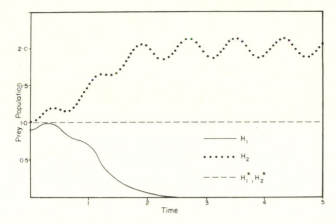

FIGURE 3.4. Behavior of the prey populations, $H_1$ and $H_2$, in another particular example of the general equation (3.4). Here the initial conditions are $H_1(0) = 0.9$, $H_2(0) = P_1(0) = P_2(0) = 1.0$, and the coefficients in equation (2.2) here have the proportions $a_1 = a_2 = 3$, $\alpha_{11} = \alpha_{22} = 1$, $\alpha_{12} = \alpha_{21} = 2$, $b_1 = 40$, $b_2 = 20$, $\beta_{11} = 3$, $\beta_{12} = \beta_{21} = \beta_{22} = 1$. The broken line again represents the "equilibrium" populations, renormalized to be unity.

As stressed by Kormondy (1969, p. 97), one feature suggested by the Lotka-Volterra predator-prey equations (3.1) has substantial practical implications for pest control strategies. If both prey and predator populations are reduced from their equilibrium values, for example by application of pesticides, the dynamics of the system leads to a consequent further reduction in the predator numbers, and an increase in the prey abundance beyond its original equilibrium value. This feature remains true in our neutrally stable $n$-predator–$n$-prey models (3.4) for any factor which reduces all populations below their equilibrium values.

The above results may be generalized in two respects.

First, although these results have been stated for the multispecies generalization of the Lotka-Volterra equations, they can be extended to the wider class of models

47

where we have $m$ interacting species, with populations $N_i(t)$, governed by $m$ equations of the special form

$$\frac{dN_i(t)}{dt} = F_i(N_i)G_i(N_1, N_2, \ldots, N_j, \ldots, N_m : j \neq i), \quad (3.8)$$

with $i = 1, 2, \ldots, m$. That is, in the growth rate of the $i$th species, the dependence on the species itself ($F_i$) can be factored out from the dependence on all the other species ($G_i$, which can be a function of all the populations $N_j$ except $N_i$ itself). Furthermore we assume that the $m$ non-zero equilibrium populations are obtained by solving the $m$ equations $G_i = 0$. This factoring condition is admittedly artificial, but apart from this the functions $F_i$ and $G_i$ can be arbitrarily complicated. The $n$-predator–$n$-prey equations (3.4) are clearly a special case of this general form (3.8), with $F(N) \equiv N$, and simple linear $G$. The assumptions imply that the diagonal elements of the community matrix are all zero: $a_{ii} = 0$. The result discussed in Appendix II (p. 195) then says that, just as for the earlier model (3.4), disturbances from equilibrium here in (3.8) lead at best to purely oscillatory behavior (when all the community matrix eigenvalues lie on the imaginary axis, having real parts zero). More usually the equilibrium is an unstable one, in the sense of Figure 2.1(a), and the system purges itself of species until it is simplified to a neutrally stable one. For further details, see May (1971).

Second, returning to our model (3.4), we may remove the restriction that the number of prey and predator species be equal. Suppose we have $m$ predator and $n$ prey species, with for definiteness $m > n$, say $m = n + q$. We now have $n + q$ equations for the $n$ variables $H_i^*$, and it must be that only $n$ of these equations are linearly independent, with constraints on the coefficients $b_i$ and $\beta_{ij}$ ensuring that the other $q$ equations are consistently redundant. For the $n + q$ variables $P_i^*$ we have only $n$ equations, which means that $q$ of the $P_i^*$ can be chosen arbitrarily, and the remaining $n$ then determined (subject overall to the condition that all $P_i^*$ be positive). The consequent stability analysis yields $2n$ eigenvalues occurring in $n$ pairs exactly as in the $n$-predator–$n$-prey system, along with a further $q$ eigenvalues which are all identically zero, corresponding to the $q$ predator populations which can be varied arbitrarily. Thus the result above is unchanged in essentials.

The consistency relations necessary to have $q$ redundant equations may be argued to be "infinitely unlikely" in nature. Such a viewpoint, which may be put in algebraic form (as here), or in geometric or topological form (e.g. it is infinitely unlikely for three planes to intersect along one line), is the basis of the celebrated "number of species equals number of resources" theorem of MacArthur and Levins (1964), Levins (1968a), Rescigno and Richardson (1965), and recently elaborated by Levin (1970). In the present context, this theorem would say that necessarily $m = n$. Although possessing undoubted elegance and affording insights in certain circumstances, this theorem is often tautological in application. When one finds a system where the number of predators exceeds the number of prey species (their resources), one observes that different utilizations are made of the prey by different predators, so that one prey species may represent more than one resource. This way of defining away the difficulty is not very different from regarding the "infinitely unlikely" constraints as being in fact fulfilled.

In conclusion, we emphasize that whether or not the Lotka-Volterra equations are applicable to real-world situations is beside the point being made here, which is that simple mathematical models with many species are in general less stable than the corresponding simple mathematical models with few species.

## MODELS WITH SPECIAL SYMMETRIES

There is a body of literature dealing with multispecies Lotka-Volterra models with special symmetries among the interaction parameters. In, passing, it is well to review these. We first discuss such models for predator-prey systems, and then for competitors.

### Predator-Prey Equations with Antisymmetry

A much discussed generalization of the Lotka-Volterra equations to $m$ interacting species, with populations $N_i(t)$, is †

$$\frac{dN_i(t)}{dt} = N_i(t)[a_i - \sum_{j=1}^{m} \alpha_{ij}N_j(t)], \qquad (3.9)$$

where the interaction coefficients $\alpha_{ij}$ are antisymmetric: $\alpha_{ij} = -\alpha_{ji}$ (and so the diagonal elements are zero, $\alpha_{ii} = 0$).

Obviously this system is a quite special case of the general form (3.8) discussed above. It is related to, but more special than, our $n$-predator–$n$-prey equation (3.4). In (3.9), if the coefficient $\alpha_{ij}$ is positive, the $j$th species preys upon the $i$th, and conversely, if $\alpha_{ij}$ is negative, the $j$th is preyed upon by the $i$th, so that a given species can act as prey and as predator in its several interactions. Our model (3.4) is more restricted in that clear distinction is made between prey species and predator species; there are but two trophic levels (cf. Figure 3). However the antisymmetry confers very special properties on the system (3.9). In our model (3.4) we would recover the essentials of (3.9) if we were to demand $\beta_{ij} = \alpha_{ji}$.

The possible equilibrium populations, $N_i^*$, are obtained in the usual way by setting the terms in square brackets in equation (3.9) equal to zero:

† Kerner's (1957, 1959, 1969) version of this equation has an additional parameter $\beta_i$. Replacing Kerner's $N_i$ by $N_i/\beta_i$, and his $a_{ij}$ by $\beta_i\beta_j\alpha_{ij}$, leads to the canonical form (3.9).

50

$$a_i = \sum_{j=1}^{m} \alpha_{ij} N_j^*. \qquad (3.10)$$

As first noted by Volterra, a consequence of the antisymmetry is that the system (3.9) exhibits purely oscillatory behavior when displaced from equilibrium. This is because the eigenvalues of an antisymmetric matrix are necessarily all purely imaginary numbers, corresponding to neutral stability. (Although the community matrix has elements $a_{ij} = -N_i^* \alpha_{ij}$, rather than simply $-\alpha_{ij}$, the theorem still applies: see Appendix II, p. 194.) This underlines the special nature of this system, and is to be contrasted with the more general results obtained from our multispecies equation (3.4) of the preceding section (which already is special enough).

Physically, the magnitude of the coefficient $\alpha_{ij}$ can be regarded as a measure of the predator's efficiency, and the antisymmetry expresses the assumption that the predator populations' gains are all directly proportional to the prey populations' losses. By a more explicit discussion, along the lines laid down by MacArthur (1969, 1970), we can show the antisymmetry to follow if the biochemical conversion efficiency of one gram of a resource (say the prey species labeled $j$) is a constant $C_i$ for members of the $i$th predator species, quite independent of which prey species is being eaten, i.e. independent of $j$. This assumption is open to criticism, and Kerner's (1969) and Leigh's (1965, 1968) suggestion that antisymmetry may represent a good approximation, with realistic corrections tending to have canceling effects, although appealing seems to have little empirical foundation.

Given the system (3.9), and the equilibrium populations $N_i^*$, Kerner (1957) showed that the quantity

$$\Phi \equiv \sum_{i=1}^{m} \{N_i(t) - N_i^* \ln N_i(t)\} \qquad (3.11)$$

is conserved. This may be verified by differentiating (3.11) and using equations (3.9) and (3.10) to get

$$\frac{d\Phi}{dt} = - \sum_{i,j=1}^{m} (N_i - N_i^*)\alpha_{ij}(N_j - N_j^*). \qquad (3.12)$$

The quadratic form on the right-hand side automatically vanishes if the matrix $\boldsymbol{\alpha}$ is antisymmetric, establishing the conservation law that $\Phi$ remains constant as the populations $N_i(t)$ vary according to equation (3.9).

This conservation law is associated with the purely oscillatory, neutral stability of the system. In just such a way is the conservation of mechanical energy associated with the neutral stability of the frictionless pendulum. Once such a constant of the motion has been found, one can invoke from physics the whole machinery of statistical mechanics. Such a program, as carried out by Kerner (1957, 1959, 1969) and Leigh (1968, 1971), and most recently efflorescing in the work of Goel, Maitra, and Montroll (1971), constructs a "temperature" which measures the average deviation of population numbers from their average values $N_i^*$, an equipartition theorem, an expression for flow of "heat" between weakly coupled systems, and so on. For lucid reviews, see Kerner (1969) or Leigh (1971).

These results are undeniably very elegant, but they are fragile, as they ultimately rest on the precise antisymmetry of the $\alpha_{ij}$ coefficients; the underlying conservation law (3.11) is much less robust than the spatial and temporal invariance of the laws of physics which underpin conventional statistical mechanics. As emphasized by Lewontin (1969), the model itself is structurally unstable.

A disquieting point, first remarked by Volterra (1931, 1937), and acknowledged by all others since, is that the models apply only to systems with an even number of species. They relate to a 60-species community, but not to a 61-species one. Nor is this some minor peripheral feature. The antisymmetry implies that the community matrix eigenvalues occur in complex conjugate pairs on the imaginary axis: this lack of damping or growth is intimately associated with the central conservation law; but equally it implies there must be an even number of species (otherwise the odd, unpaired, eigenvalue must necessarily be zero, leading to a bothersome singularity in the community matrix). Thus the conservation law (3.11) and the restriction to an even number of species in the community have the same truth content. Some people are untroubled by this. Kerner (1961) has discussed in detail the dynamical process whereby a community with an odd number of species collapses to an even-numbered one, and in particular how a 3-species such system collapses to a 2-species one. Deakin (1971) has exhaustively catalogued all the topologically distinct food web structures compatible with (3.9) for the communities with $m = 2, 4, 6, 8, 10$ species. My view is that such constraints border on the ridiculous, and highlight the unwholesomeness of the models.

In short, I am skeptical of any interpretation of the fluctuations observed in natural populations which is based on the pathological neutral stability character of a set of specially antisymmetric Lotka-Volterra equations.

A more technical criticism is that if the system is too richly connected, in the sense that too many of the species interact with each other (too many of the $\alpha_{ij}$ are non-zero), the methods of statistical mechanics are inapplicable because the ergodic hypothesis — a particular sort of randomness assumption — fails. This point, raised by Goel (see the review in Rosen, 1970, p. 296), not only cuts at Kerner's and Leigh's work, but again reminds us that too much complexity in our multispecies system is not necessarily helpful.

## Competition Equations with Symmetry

Another interesting system is that for a single trophic level having $m$ competitors, with populations $N_i$:

$$\frac{dN_i(t)}{dt} = N_i(t)[a_i - \sum_{j=1}^{m} \alpha_{ij} N_j(t)], \qquad (3.13)$$

with † $i = 1, 2, \ldots, m$. Here it is required that all $a_i$ are positive, and all $\alpha_{ij}$ are, at least, not negative. Unlike the earlier models defined by equations (3.4), (3.9), or even (3.8), these equations contain explicit intraspecific interactions. The self-interaction coefficients $\alpha_{ii}$ are no longer zero. Thus, as remarked earlier (p. 42), it can now be said without reservation that, although the equations (3.13) are unrealistically simple, they do characterize the stability properties of a wider class of models, to which they are the first approximation in a Taylor series.

So far the system (3.13) is rather general. By making the special *symmetry* assumption $\alpha_{ij} = \alpha_{ji}$, MacArthur (1970) showed that the quadratic form

$$Q(t) = \sum_{i,j=1}^{m} (N_i(t) - N_i^*)\alpha_{ij}(N_j(t) - N_j^*) \qquad (3.14)$$

is minimized by competition. That is,

$$\frac{dQ(t)}{dt} \leq 0. \qquad (3.15)$$

Here the $N_i^*$ are, as usual, the equilibrium populations, given again by equation (3.10). I have rewritten MacArthur's $Q$ in an equivalent and more manifestly symmetric form, which will be helpful below.

The proof of the assertion (3.15) proceeds by differen-

---

† Actually, MacArthur's (1970) equation contains an extra factor $C_i$. The canonical form (3.13) can be recovered by replacing MacArthur's $\chi_i$ by $C_i N_i$, and his $\alpha_{ij}$ by $\alpha_{ij}/C_i C_j$.

tiating equation (3.14) for $Q$, and using the basic dynamical equation (3.13). Further using the relation (3.10) to express the $a_i$ in terms of a sum over the coefficients $\alpha_{ij}$ and the populations $N_j^*$, we get

$$\frac{dQ}{dt} = -\sum_{i,j,k} [N_i\alpha_{ik}(N_k - N_k^*)\alpha_{ij}(N_j - N_j^*)$$

$$+ (N_i - N_i^*)\alpha_{ij}N_j\alpha_{jk}(N_k - N_k^*)]. \quad (3.16)$$

In view of the symmetry property $\alpha_{ij} = \alpha_{ji}$, this may be rewritten as

$$\frac{dQ(t)}{dt} = -2\sum_{i=1}^{m} N_i(t)[J_i(t)]^2, \quad (3.17)$$

where we have defined

$$J_i(t) = \sum_{k=1}^{m} \alpha_{ik}(N_k(t) - N_k^*). \quad (3.18)$$

Since all the populations must be positive numbers, and each squared quantity $J_i^2$ must be positive, the result (3.15) follows. The equality $dQ/dt = 0$ obviously pertains if and only if all $J_i$ are zero, which is to say at the equilibrium point where all populations have the values $N_i^*$.

This result is especially interesting if, as is usually the case, the matrix of competition coefficients $\alpha_{ij}$ is positive definite (i.e. if all its eigenvalues, which are necessarily real because the matrix is symmetric, are positive). In this event, the quadratic form $Q$ is necessarily also positive definite,

$$Q(t) > 0 \quad (3.19)$$

for all population values, with the exception of the equilibrium point where $Q = 0$. The properties (3.19) and (3.15) constitute the definition of a Lyapunov function, as discussed earlier. Consequently the full nonlinear global stability analysis of the system is here legitimately described by the linearized neighborhood analysis. This is a useful fact.

In pictorial terms, "Using $N_1$, $N_2$, . . . as coordinates, we can plot $Q$ as a landscape whose primary feature is a regular valley. A point on this landscape is a set of populations of $N_i$ and our theorems say that as competition acts, the point moves down the landscape, the equilibrium point being the lowest point of the valley. If the bottom of the valley has some negative $N$ coordinates, those species are eliminated" (MacArthur, 1970, p. 10).

Under certain restrictions, a physical interpretation of $Q$ can be given, as the square on the deviation of available production from the species' intrinsic harvesting abilities. We shall develop this further in Chapter 6.

All these results hinge on the special symmetry assumption. As mentioned above, MacArthur's physical justification can be seen to rest on the postulate that the biochemical efficiency of conversion of grams of any resource (labeled $j$, say) into grams of the population $N_i$ is a constant $C_i$, which depends only on the consuming species $i$, and not on the different resources. In general one might expect this factor to involve both $i$ and $j$: $C_{ij}$.

In the work on the predator-prey equations (3.9), the special antisymmetry is absolutely essential; if it is broken by the smallest amount, the conservation law (3.11) is lost, and with it all the consequent theoretical edifice. What happens if the exact symmetry is broken in the competition equations (3.13)?

It can be shown that the expression $Q(t)$ remains minimized by competition, that is equation (3.15) remains valid, if the symmetry of the competition coefficients is violated by an amount small compared to the smallest eigenvalue of the $\alpha$ matrix. That is, we can tolerate a symmetry breaking such that $\alpha_{ji} = \alpha_{ij} + \epsilon_{ij}$, provided the quantities $\epsilon_{ij}$ are typically small in this sense. The result uses the fact that the eigenvalues $\lambda(A')$ of a matrix $A' = A + \epsilon B$ differ smoothly from the eigenvalues $\lambda(A)$ of the original matrix $A$ by amounts of the order of $\epsilon$, where $\epsilon$ is by assumption small.

In summary, the minimum principle expressed by equations (3.14) and (3.15) is an elegant result, particularly in that it can justify the congruence between neighborhood and global stability properties for these model ecosystems. Although fragile in requiring that the competition coefficients be symmetric, MacArthur's work is not endowed with quite the same fragility as the corresponding antisymmetric predator-prey models. The essential difference between the structurally unstable results for the antisymmetric case, and the structurally stable results for the symmetric models, is that the former require all the eigenvalues of the community matrix to lie exactly on the imaginary axis, whereas the latter have no such extreme requirement.

Toward the end of his paper, MacArthur (1970, Sect. 5) notes that the symmetry in his model has one remarkable consequence, which follows from the mathematical fact that if an $m \times m$ symmetric matrix is augmented into an $(m + 1) \times (m + 1)$ one by addition of an extra row and column, the smallest eigenvalue of the augmented matrix is less than, or equal to, that of the $m \times m$ matrix. Thus, in MacArthur's words, "when an additional species is added to a community of competitors with symmetric $\alpha$ matrix, the stability as measured by the smallest eigenvalue cannot increase and usually decreases." Actually, the matter is not as simple as put in MacArthur's paper, since the stability behavior depends not only on the $\alpha$ matrix but also on the equilibrium populations $N_i^*$; that is the stability-setting community matrix has elements $a_{ij}$ which are not simply $-\alpha_{ij}$, but rather are $-N_i^*\alpha_{ij}$. (An extreme example of this point is provided by equation (3.3) for the two-species

Lotka-Volterra model, where the eigenvalues manifestly involve only $a$-type parameters, and not $\alpha$-type coefficients.) However, a more careful consideration seems to preserve MacArthur's conclusion intact in essentials. The upshot is that within this symmetric model not only is the general $(m + 1)$-species community less stable than an $m$-species one, but in particular any single species by itself is more stable (i.e. has a quicker damping time) than any two-species assembly, and so on. This theorem again sounds our theme: the more the species in this single trophic level model, the less the stability.

## STABILITY AT ONE TROPHIC LEVEL VERSUS WEB STABILITY

Before moving on to more general kinds of model, we pause to take up the question of the relation between stability within any single trophic level and stability of the total trophic web. In an effort to codify the often confusing field results mentioned at the start of this chapter, various conjectures have been advanced.

Watt (1968, Ch. 3.3) and Zwolfer (1971) have held that stability at one particular trophic level (e.g. competitive stability among predators) may promote overall instability (e.g. an herbivore species escapes control).

Conversely, Paine (1966) has elaborated a conjecture of Hutchinson (1961) that, although one particular trophic level may in isolation be unstable due to competition, the effects of other levels (e.g. predation) can lead to a total system which is stable. This view has received support from Hall, Cooper, and Werner (1970), in whose freshwater communities the bluefish plays a dominant predator role akin to Paine's *Pisaster*, and from Connell (1971), who goes beyond his field studies on intertidal barnacles and rain forest trees to make tentative suggestions as to the environmental conditions under which the Paine mechanism is likely to operate.

To the contrary of both these viewpoints, Holling (1968b) has argued that stability at any one trophic level should, by feedback, encourage stability at other levels, and thence of the total web. The intuition of the physical scientist would support this view and the corollary that instability at any one level should tend to create an unstable total system.

In a review of the question, Southwood and Way (1970) emphasize that there is unlikely to be a general answer which is not bound up with the topological details of the particular web under consideration.

Our mathematical models can make a contribution to this discussion.

To this end, consider the simple 2-predator–2-prey community obtained from equation (3.4) by adding explicit competition between the two prey populations:

$$\frac{dH_i(t)}{dt} = H_i(t)[a_i - \sum \epsilon_{ij}H_j(t) - \sum \alpha_{ij}P_j(t)] \quad \text{(3.20a)}$$

$$\frac{dP_i(t)}{dt} = P_i(t)[-b_i + \sum \beta_{ij}H_j(t)]. \quad \text{(3.20b)}$$

Here the indices $i$ and $j$ take the values 1 and 2, and the coefficients $\epsilon_{ij}$ measure the direct competition between and within the prey populations. In a like manner, we could alternatively introduce direct competition within the predator trophic level.

First suppose both predator populations are constrained to be zero, so that the community comprises only the two competing prey species. The criterion for this single trophic level system to be stable is the familiar

$$\epsilon_{11}\epsilon_{22} - \epsilon_{12}\epsilon_{21} > 0. \quad \text{(3.21)}$$

Equation (3.21) tends to require that each species encounters greater competitive stress from its own, rather than from the other, species.

To study the stability of the full four-species web whose dynamics are given by equation (3.20), it is necessary to solve a quartic equation for the four eigenvalues of the community matrix. It is obvious that the stability criterion for the full system will not be identical with the simple criterion (3.21). Even so, it may be seen (May, 1971) that this more complicated stability condition usually does manifest certain similarities with (3.21), in that relatively large values of $\epsilon_{11}$ or $\epsilon_{22}$ make for stability, and relatively large values of $\epsilon_{12}$ or $\epsilon_{21}$ for instability, of the total system. To comprehend the structure of the stability criterion for the four-species system, we take the pathologically simple case of "equal predation" (Parrish and Saila, 1970), where the stability criterion can be shown to have the specially simple form (May, 1971)

$$\epsilon_{11} + \epsilon_{22} - \epsilon_{12} - \epsilon_{21} > 0. \qquad (3.22)$$

We note that (3.21) and (3.22) are by no means identical, but that random choices of the four parameters will in general tend to satisfy, or not to satisfy, both equations together. Try it. However, as an extreme example, if $\epsilon_{11} = 0$ and $\epsilon_{22} \gg \epsilon_{12}, \epsilon_{21}$, then (3.22) is satisfied and (3.21) is not, exhibiting the behavior conjectured by Paine. Conversely, if $\epsilon_{12} = 0$ but $\epsilon_{21} \gg \epsilon_{11}, \epsilon_{22}$, (3.21) is satisfied but (3.22) is not, exemplifying Watt's conjecture.

In this context, it is instructive to review the work of Parrish and Saila (1970). Motivated by Paine's conjecture, they consider a 1-predator–2-prey system governed by the analogues of equations (3.20) with but one predator population. Beginning with prey systems competitively unstable in the absence of predation, they carry out computations to see whether including predation can stabilize the total system. However, the numerical coefficients they happen to choose always leave the total system unstable, although the doomed prey species often persists much longer than in the absence of predation. Thus Parrish and Saila's computer experiments do not in fact yield an example illustrating the working of Paine's conjecture, but rather bear eloquent witness to the general tendency for single level stability and total system stability to go hand in hand. An analytic investigation of Parrish and Saila's model

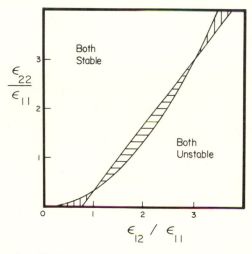

FIGURE 3.5. Illustrating how stability within one trophic level in isolation (here prey stability with respect to competition in the absence of predators, as measured by the criterion (3.21)) is related to total trophic web stability (here, for the very special case of "equal predation," measured by the criterion (3.22)), in the two-predator—two-prey system (3.20).

The stability criteria involve the parameter ratios $\epsilon_{22}/\epsilon_{11}$ (the $y$ axis), $\epsilon_{12}/\epsilon_{11}$ (the $x$ axis), and $\epsilon_{21}/\epsilon_{11}$ (the $z$ axis, not shown): a typical 2-dimensional slice across this 3-dimensional parameter space is shown, namely the fixed value $\epsilon_{12} = 3\epsilon_{21}$. The upper left region corresponds to both single level and total web being stable; the lower right region to both unstable. The small, horizontally hatched region in the center corresponds to single-level intrinsic stability along with total web instability (Watt's conjecture), while the two small vertically hatched regions have single-level instability redeemed by total system stability (Paine's conjecture).

6 1

(Cramer and May, 1971) shows that in the event of "equal predation," which is considered throughout their paper, the criterion for total system stability is again (3.22). Armed with this insight, one may seek out the relatively small corner of parameter space where indeed the condition (3.21) is violated but (3.22) is satisfied, thus illustrating Paine's idea.

These remarks are borne out by Figure 3.5, which shows the relation between single level stability (according to the criterion (3.21)) and total system stability (according to the very special criterion (3.22), which tends to characterize more general total systems), for various values of the competition parameters. To keep the figure two-dimensional, we put $\epsilon_{12} = 3\epsilon_{21}$.

In short, these general mathematical models point the way to a synthesis of the views outlined above. We see that the criteria for stability at any one level and for the total system are not identical. However, the criteria tend to be similar, so that usually stability at one trophic level tends to go along with stability of the total web, and conversely with instability. Nevertheless, by judicious choice of the various competition and interaction parameters, it can be arranged that one level would in isolation be stable, but that the total system is unstable (Watt's conjecture); or, alternatively, it can be arranged that, although one level would in isolation be unstable with respect to competition, the total web is stable (Paine's conjecture). Whether natural evolutionary processes seek out these special corners of parameter space is another question.

## RANDOMLY CONSTRUCTED WEBS

We now move on to take a more abstract view of the community of interacting populations. Rather than deal with some actual set of differential equations for the population dynamics, we define and discuss the model ecosystems in terms of the community matrix $A$. As emphasized in Chapter 2, assumptions as to the biological construction of the food web correspond to assumptions about the structure of $A$. In turn, the eigenvalues of $A$ tell us the stability consequences of these biological assumptions.

Recently Gardner and Ashby (1970) have studied the stability properties of large complex systems whose component elements are connected at random. Their conclusion, based on the trend of computer studies of systems with 4, 7, and 10 variables, is that such systems may be expected to be stable up to some critical level of connectance, and beyond this point to go suddenly unstable.

We proceed to elaborate this work. Following Gardner and Ashby, suppose we have a community comprising $m$ species, each of which would by itself have a density-dependent or otherwise stabilized form, so that if disturbed from equilibrium it would return with some characteristic damping time. To set a time-scale, these damping times are all chosen to be unity; that is each population in isolation would contribute a community matrix element $a_{ii} = -1$. Next we "switch on" the interactions. The web connectance, $C$, expresses the probability that any pair of species will in fact interact. It is measured as the percentage of non-zero elements in the matrix, or equivalently as the ratio of actual links to topologically possible links in the trophic web. Consequently, of the matrix elements $b_{ij}$ which are switched on by the interactions, a fraction $1 - C$ remain zero.

With probability $1 - C$: $\quad b_{ij} = 0.$ \hfill (3.23)

The remaining non-zero interaction elements are each assumed equally likely to be positive or negative, having an absolute magnitude chosen from some statistical distribution. That is, each of these matrix elements is assigned from a distribution of random numbers, which distribution itself has mean value zero, and mean square value $s^2$. $s$ may be thought of as expressing the average interaction "strength," which average is for simplicity common to all interactions.

$$\text{With probability } C: \quad b_{ij} = \begin{bmatrix} \text{chosen from random} \\ \text{number distribution,} \\ \text{mean} = 0, \text{ mean} \\ \text{square} = s^2. \end{bmatrix} \quad (3.24)$$

In short, the final community matrix for the system randomly assembled in this fashion is

$$A = B - I. \qquad (3.25)$$

The $m \times m$ random matrix $B$ is as defined by equations (3.23) and (3.24), and $-I$ represents the intrinsic negative feedback assumed for each population, with $I$ being the $m \times m$ unit matrix. We thus have an unbounded ensemble of models, one for each specific choice of the interaction matrix elements drawn individually from the random number distribution.

It is important to note that the randomness only enters in the initial choice of the coefficients $b_{ij}$, which then define a particular model. Once the dice have been rolled to get a specific system, the subsequent analysis is strictly deterministic.

According to Figure 2.2, the system corresponding to the community matrix $A$ is stable if and only if all the eigenvalues of $A$ have negative real parts. For a system with a specified number of species $m$, connectance $C$, and average interaction strength $s$, it is interesting to ask what is the probability $P(m, C, s)$ that a particular matrix drawn from the ensemble will correspond to a stable community.

For large $m$, analytic techniques developed for treating large random matrices may be used to show that the community matrix $A$ will be almost certainly stable, $P(m, C, s) \rightarrow$ 1, if (May, 1972b)

$$s(mC)^{1/2} < 1, \tag{3.26}$$

and almost certainly unstable, $P \rightarrow 0$, if

$$s(mC)^{1/2} > 1. \tag{3.27}$$

The transition from stability to instability as $s$ increases from the regime (3.26) into the regime (3.27) is very sharp for $m \gg 1$. (Plausibility arguments suggest the relative width of the transition region scales as $m^{-2/3}$.) The central feature of these results for large systems is the sharp transition from stable to unstable behavior as either the number of species $m$, or the connectance $C$, or the average interaction strength $s$, exceeds a critical value.

Such a definite answer for any model in the ensemble in the limit $m \gg 1$ is a consequence of the familiar statistical fact that, although individual matrix elements are liable

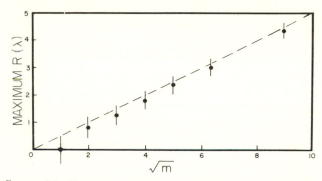

FIGURE 3.6. The asymptotic ($m \gg 1$) analytic result embodied in equations (3.26) and (3.27), and shown by the dashed line here, is contrasted with numerical Monte Carlo studies of the dominant eigenvalue of random matrices of various sizes. We fix the root-mean-square interaction strength $s = 0.5$, and the connectance $C = 1$, to show the largest real part, $R(\lambda)$, of eigenvalues of matrices drawn from such ensembles. The mean values and standard deviations coming from such a computer study are shown as a function of $m^{1/2}$ for various $m$.

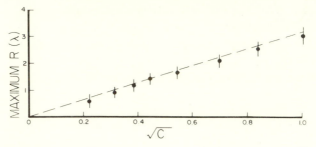

FIGURE 3.7. As in Figure 3.6, except now in the computer studies we fix $s = 0.5$, the number of species $m = 40$, and vary the connectance $C$. The numerical results for the largest $R(\lambda)$ are displayed as a function of $C^{1/2}$; again the dashed line is the asymptotic analytic approximation, as in equations (3.26) and (3.27).

to have any value, by the time we have an $m \times m$ matrix with essentially $m^2$ such statistical elements, the total system will have relatively well-defined properties.

These conclusions come from the trivial observation that the eigenvalues of $A$ are $\lambda(A) = \lambda(B) - 1$, together with the result that the largest real part of the eigenvalues of the random matrix $B$ is asymptotically $s(mC)^{1/2}$. The analytic result can be buttressed with Monte Carlo computer studies of the eigenvalue spectrum of random real matrices (McMurtrie, 1972). In Figure 3.6 we fix $s = 0.5$ and $C = 1$, and show numerical results for the largest eigenvalues of matrices drawn from such ensembles, for various $m$. Figure 3.7 shows like results for fixed $s = 0.5$, $m = 40$, and varying connectance $C$. Although our analytic results are based on the assumption that $m \gg 1$, and are therefore only approximate when applied to moderate $m$, they agree well with the numerical experiments.

In brief, this ensemble of very general mathematical models of multispecies communities, in which the population of each species would by itself be stable, displays the property that too rich a web connectance (too large a $C$) or too large an average interaction strength (too large an

$s$) leads to instability. The larger the number of species, the more pronounced the effect is.

Two corollaries are amusing, although they should not be taken too seriously.

First, notice that two different systems of this kind, with average interaction strengths and connectances $s_1$, $C_1$ and $s_2$, $C_2$, respectively, have similar stability character if (see equation (3.26))

$$s_1^2 C_1 \simeq s_2^2 C_2.$$

Roughly speaking, this suggests that, within a web, species which interact with many others (large $C$) should do so weakly (small $s$), and conversely those which interact strongly should do so with but a few species. This is indeed a tendency in many natural ecosystems, as noted for example by Margalef (1968, p. 7): "From empirical evidence it seems that species that interact feebly with others do so with a great number of other species. Conversely, species with strong interactions are often part of a system with a small number of species. . . ."

A second feature of the models may be illustrated by using Gardner's and Ashby's computations (which are for a particular interaction strength) to see, for example, that 12-species communities with 15% connectance have a probability essentially zero of being stable, whereas if the interactions be organized into three separate $4 \times 4$ blocks of 4-species communities, each with a consequent 45% connectance, the "organized" 12-species models will be stable with probability 35%. That is, of the infinite ensemble of these particular 12-species models, essentially none of the general ones is stable, whereas 35% of those arranged into three "blocks" are stable. Such examples suggest that our model multispecies communities, with given average interaction strength and web connectance, may do better if the interactions tend to be concentrated in small blocks, rather than distributed uniformly throughout the web—again a feature observed in many natural ecosystems.

## MISCELLANEOUS OTHER MODELS

In a most interesting if difficult paper, quite independent of the work covered so far in this chapter, Levins (1970b) has observed that: "A number of quite independent lines of argument converge toward the assertion that there is often a limit to the complexity of systems."

One of the main such arguments draws upon Kauffman's (1970a, b) work on large randomly constructed nets (see also, Slone, 1967, and Burns, 1970). This analytic and computational work, which is still in a developing state, is motivated toward understanding how metabolic stability emerges from nets of interactions between thousands, or even millions, of chemical species. The emergent mathematical insights may, however, be bent to ecological purposes. The principal feature of these randomly assembled, deterministic nets is that sensible results are found only for nets with low connectance. In nets where each element receives inputs from too many other elements, the system exhibits physically unreasonable state cycles with times rapidly becoming astronomically large; the existence of longish loops in the structure also causes instability. This obviously supports Levins' (and our) thesis.

Levins also refers to the associated, but more qualitative, work of Ashby (1952), who in his design for a brain argues for loosely coupled subsystems as optimal.

Another of Levins' lines of argument rests on the stability properties of general systems, whose structure is characterized by a matrix $a_{ij}$. This work, although different in detail, is in the same spirit as that of the preceding section. Levins' conclusions are essentially based on the sufficient stability criterion $s < m^{-1}$, which is much weaker than the

FIGURE 3.8. A schematic representation of a simple trophic web with three levels. Such a web may be visualized as a plant species (1), an herbivore species (2), and an omnivore (3), which eats both (1) and (2).

full necessary-and-sufficient condition (3.26), but points in the right direction, namely that too many strong links will lead to a system which is almost certainly unstable. In other words (Levins, 1970, p. 86): "The dynamics of a broad class of complex systems will result in simplification through instability."

In his interesting paper on the systems analysis of energy flow and population stability in relatively simple food webs, Hubble (1973) draws a parallel moral. Hubble considers an example which, in essence, corresponds to the trophic web of Figure 3.8, where a prey species (2) and a predator species (3) also compete for a common resource (1). He finds the system can be unstable, but that all potentially unstable cases can be stabilized by cutting either the trophic link between 2 and 3 or that between 3 and 1. As Hubble says, "This simple example . . . illustrates that the trophic complexity of a food web need bear no necessary relationship to its dynamic stability—a relationship that has often been asserted to exist."

Yet again, Smith (1969) has considered models based on

simple 3- or 4-species straight chain ecosystems. If the initial simple system is stable, additional complication and diversity can usually be grafted on without destroying the stability. On the other hand, if the basic simple system is unstable, the addition of complications and diversity generally makes things even worse.

## QUALITATIVE STABILITY

In one form or another, all the work discussed so far makes assumptions about the *magnitudes* of the interactions between species in the community, that is, about the magnitude of the elements in the community matrix.

What can be said, knowing only the *signs* of the individual matrix elements (+, −, or 0), and nothing else?

In general, if a matrix can be shown to be necessarily stable (i.e. all eigenvalues having negative real parts), altogether independent of the actual magnitude of the non-zero elements, the matrix is called "qualitatively stable." This is an important subject in mathematical economics, where often no quantitative information is available (Quirk and Ruppert, 1965: Maybee and Quirk, 1969).

The situation in ecology is similar. The sign of the community matrix elements can often be found by inspecting the food web diagram, even in the total absence of any quantitative data. This intimate association between the sign structure of the community matrix and the qualitative nature of the biological interactions was emphasized earlier (p. 25).

The necessary and sufficient conditions for a matrix to be qualitatively stable are set out below. If the signs (+, −, or 0) of the various matrix elements satisfy these detailed criteria, then the system is stable. If the criteria are not obeyed, nothing can be said: the matrix may be stable or unstable, depending on the actual magnitudes of the matrix elements. Usually this set of mathematically rigorous qualitative stability criteria will not apply exactly to complicated real-world situations, but they still help to indicate general trends.

In mathematical terms, the necessary and sufficient conditions for an $m \times m$ matrix $A$, with elements $a_{ij}$, to be qualitatively stable are (Quirk and Ruppert, 1965):

(i)    $a_{ii} \leq 0$, all $i$.                               (3.28)

(ii)   $a_{ii} \neq 0$, at least one $i$.               (3.29)

(iii) $a_{ij}a_{ji} \leq 0$, all $i \neq j$.                 (3.30)

(iv) For any sequence of 3 or more indices $i$, $j, k, \ldots, q, r$ (with $i \neq j \neq k \neq \ldots \neq q \neq r$), the product $a_{ij}a_{jk} \ldots a_{qr}a_{ri} = 0$.      (3.31)

(v)   $\det A \neq 0$.                                     (3.32)

It is worth restating the qualitative stability criteria (i) − (v) in biological terms.

The first two conditions pertain to intraspecific effects. Condition (i) requires that no population exhibit a destabilizing positive feedback in its intraspecific interactions, and condition (ii) further demands that at least one population in the community actually exhibit a self-stabilizing effect. Condition (iii) has the consequences discussed more fully below; that symbiotic relations (++) have the same qualitative stability character as competitive ones (−−) may not be intuitively obvious. Condition (iv) forbids closed loops of three or more members, in the sense that the effects of $i$ on $r$, $r$ on $q$, $\ldots$, $k$ on $j$, and finally $j$ back on $i$ are all non-zero. Without the trivial overriding condition (v), requiring $A$ to be non-singular, the system would be under-determined; there would in effect be more populations than there were equations, and one or more populations could be assigned arbitrary values.

As an illustration of these ideas, we consider the food web of Figure 3.8, corresponding to Hubble's example

71

discussed above. If we suppose that resource, prey, and predator all exhibit intraspecific negative feedback effects, the community matrix for this system clearly has the sign structure

$$A_1 = \begin{pmatrix} - & - & - \\ + & - & - \\ + & + & - \end{pmatrix}. \tag{3.33}$$

The matrix fails to satisfy condition (iv), and therefore is not qualitatively stable. $A_1$ may or may not correspond to a stable equilibrium, depending on the detailed numerical magnitudes of the various matrix elements. However, if we remove the link between predator and resource (the link between 3 and 1 in Figure 3.8), the simpler community matrix has signs

$$A_2 = \begin{pmatrix} - & - & 0 \\ + & - & - \\ 0 & + & - \end{pmatrix}. \tag{3.34}$$

This matrix satisfies all the conditions catalogued above, and therefore is qualitatively stable. A like result holds if the link between 2 and 3 is removed in Figure 3.8. These qualitative stability insights march with Hubble's more detailed analysis. As he emphasizes, the conclusion that the relatively simple web is generally more stable than the relatively complex one is at variance with the conventional wisdom. The example forcefully drives home the argument of Southwood and Way (1970), and others, that one must be cautious in making generalizations about the relation between population stability and the number of links in the food web structure.

It is easy to verify that any open, straight chain ecosystem, with but one population at each trophic level and no links between nonadjacent levels, will be qualitatively stable so long as there are no destabilizing intraspecific positive feedback effects, and at least one negative intraspecific feedback. The above qualitative stability conditions can bypass much cumbersome algebra in those relatively simple circumstances where they do pertain; one may thus simplify some of the work, e.g., of Bulgakova (1968 a, b) or Garfinkel (1967). However, most large natural webs will obviously violate both (iii) and (iv), so that an analysis of their stability properties requires the interaction magnitudes to be taken into account. Even so, the general tendencies revealed by qualitative stability theory are useful.

Particularly worthy of remark is condition (iii), which says that commensal $(0+)$, amensal $(0-)$, and general predator-prey $(+-)$ interactions are consistent with qualitative stability, whereas symbiotic or mutual $(++)$ and competitive $(--)$ interactions are not. This mathematically rigorous statement may be plausibly extended into the broader, if rougher, statement that competition or mutualism between two species is less conducive to overall web stability than is a predator-prey relationship. It is tempting to speculate that stability considerations may make for communities in which strong predator-prey bonds are more common than symbiotic ones. This result is not intuitively obvious, yet it is a feature of many real-world ecosystems, as observed for example by Williamson (1972, p. 95): "[Mutualism] is a fascinating biological topic, but its importance in populations in general is small." (Indeed Williamson's subsequent argument that many conventional examples of predator-prey are in fact closer to commensalism or amensalism may be pursued to suggest that not a few conventional examples of mutualism $(++)$ are in fact closer to commensalism $(0+)$. For example, in Aruego's (1970) delightful children's book *Symbiosis,* at least four of the nine pairs may be held to be commensals.)

All in all, rich trophic complexity and a diversity of different kinds of interaction between species are not conducive to qualitative stability. Insofar as the theory of qualitative stability relates to the muddied complexity-stability question, it is to re-echo the theme that, in general mathematical models, increased complexity makes for diminished stability.

## CLOSED ECOSYSTEM MODELS

An interesting subclass of systems are those which are closed, in the sense that decomposers are included, and quantities such as total biomass and total energy flux (allowing for fixation and respiration) are conserved: e.g. Demetrius (1969). No such constraints have been imposed on the systems dealt with throughout this book.

Some tentative and rather formal work on such models suggests that, on stability grounds, we may expect a tendency for the energy content per unit biomass to increase as we ascend the trophic ladder (Ulanowicz, 1972; May, 1972c). This tendency accords with the fact that animal carbohydrates and proteins have generally higher calorific values than those for plants (White, Handler, Smith, and Stetten, 1959; Morowitz, 1968).

Such work on closed ecosystem models is, however, still in a speculative and unformed stage. We shall not review it further.

## DISCUSSION

We shall not attempt to recapitulate the various minor points made in this chapter, but shall return to the main theme, namely that in general mathematical models of multispecies communities, complexity tends to beget instability rather than stability.

The oversimplified interactions between and within species embodied in the models of this chapter may be brought nearer reality in many ways. We may include destabilizing features such as time delay in responses, or saturation of predator attack capacities; or alternatively stabilizing features such as density-dependent fecundity or death rate, or predator switching. These complications can and have been introduced in the one-predator–one-prey and other few-species systems (as shall be seen in the next chapter), and it is not easy to see that anything qualitatively new will emerge when they are introduced in an analogous way into multispecies systems.

The central point remains that, if we contrast simple few-species mathematical models with the analogously simple multispecies models, the latter are in general less stable than the former. A variety of explicit counter-examples have demonstrated that a count of food web links is no guide to community stability.

This straightforward mathematical fact contradicts the intuitive verbal arguments often invoked (e.g., Hutchinson, 1959; Macfadyen, 1963, p. 182), to the effect that the greater the number of links and alternative pathways in the web, the greater the chance of absorbing environmental shocks, thus damping down incipient oscillations. The fallacy in this intuitive argument is that, the greater the size and connectance of a web, the larger the number of characteristic modes of oscillation it possesses: since in general each mode is as likely to be unstable as to be stable (unless the increased complexity is of a highly special kind), the addition of more and more modes simply increases the chance for the total web to be unstable. This is at the heart of the several general mathematical arguments reviewed above.

On the other hand, the balance of evidence would seem to suggest that, in the real world, increased complexity *is* usually associated with greater stability (e.g. Macfadyen, 1963, p. 181; Kormondy, 1969, pp. 84–111; Williams, 1964; Hutchinson, 1959; Elton, 1958; Slobodkin, Smith, and Hairston, 1967; Pianka, 1966).

There is no paradox here.

One facet of the explanation can be that, although increased complexity makes for a more unstable system, it is advantageous on other grounds; for example it may be conducive to a more thorough exploitation of the community's total resources. Then a stable environment may permit such complexity, and also be characterized by relatively unfluctuating populations. But an unstable environment may drive population instabilities which the complex system is in general ineffective in damping, with the consequence that such environments may be typified by relatively simple systems and relatively unstable populations. In short, complexity and population stability may well be associated, but no causal arrow need point from complexity to stability. To the contrary, if there is a generalization, it could be that stability permits complexity.

To have a bet each way, we add that an alternative facet of the explanation pertains to those instances where trophic complexity is indeed the agent producing population stability. There is still no contradiction with the mathematical theorem asserting this to be very unlikely in general complex systems. The real world is no general system. Nature represents a small and special part of parameter space. This point may be clarified by an analogy. Suppose a Martian were to observe those earthly laboratories where physicists are striving to confine high temperature plasmas for sufficiently long to achieve a controlled thermonuclear reactor. Our Martian would observe that simple magnetic field configurations lead to poor containment, while machines with very complicated magnetic fields generally produce more stable plasma confinement. He may be tempted to theorize that, in general, magnetic field complexity leads to stability. He would be wrong. Although simple magnetic field geometries produce unstable plasma containment, complex magnetic field configurations are almost invariably even less stable. The stable arrangements produced by complicated field geometries in plasma physics laboratories are quite atypical, having been selected with great ingenuity for this very purpose.

One particular mechanism whereby increased trophic complexity can act to help promote community stability is discussed toward the end of the next chapter.

## SUMMARY

Empirical evidence does not yet permit a decisive answer as to whether trophic richness and complexity promote population stability in the real world.

This chapter has pawed over a small piece of the puzzle. The moral emerging first from generalized multispecies Lotka-Volterra models, then from models of randomly assembled food webs, and finally from qualitative stability theory, is that, as a mathematical generality, increased multispecies trophic complexity makes for lowered stability. This fact is not inconsistent with the biological realities, but it does suggest that theoretical effort should concentrate on elucidating the very special and mathematically atypical sorts of complexity which could enhance stability, rather than seeking some (false) "complexity implies stability" general theorem.

Among other minor points, it is argued that stability (or instability) due to competition within any one trophic level usually goes with stability (or instability) of the web as a whole, although no unequivocal general statement is possible. It is seen that mutualism between species tends to have a destabilizing effect on the community dynamics Some remarks are also made about multispecies Lotka-Volterra equations with special symmetries or antisymmetries among the interaction coefficients.

# CHAPTER FOUR

# Models with Few Species:
# Limit Cycles and Time Delays

In the models just considered, all the interactions between and within species were either represented by grossly simple equations or else summarized in the vicinity of equilibrium by the elements of the community matrix. It is difficult to effect any multispecies discussion otherwise. In this chapter, attention is restricted to models with but a few species, and considerably more detail is put into the description of the dynamical interactions between populations. We first treat one-predator–one-prey models with comparatively realistic terms for the prey birth rate, the predator attack rate or functional response, and the predator population's growth rate. Then we discuss the effects of time delays in the interactions, particularly in resource limitation effects; this is done first in single-species populations, and then for systems with two trophic levels.

## COMPARATIVELY REALISTIC
## ONE-PREDATOR–ONE-PREY MODELS

The simplest predator-prey model is the Lotka-Volterra equation (3.1), which has been discussed briefly. A precise analysis of any particular real-world one-predator–one-prey situation could embrace a multiparameter systems analysis approach, but between this and the crudity of the Lotka-Volterra model lies a halfway house where one seeks to incorporate in the model a moderately realistic description of the minimum number of broadly relevant biological features. Such elements of greater realism may be introduced as follows (see also Macfadyen, 1963, Chs. 12 and 15; Clark, et al. 1967, Ch. 3; Williamson, 1972; Hassell and Rogers, 1972).

(I) The per capita birth rate of the prey population $H(t)$ (which in the Lotka-Volterra model is a constant, $a$, leading to the term $aH$ in equation (3.1a)) is in general a function of their population density. The most common formula replaces this constant $a$ with the Verhulst-Pearl logistic form

$$a \rightarrow r(1 - H/K). \qquad (4.1a)$$

As reviewed by Pielou (1969, pp. 19–21), this expression may be justified either on biological grounds (with $K$ thought of as some carrying capacity set by the environmental resources), or alternatively as the first approximation in a Taylor expansion of more general density dependences. As discussed earlier, in a single-species population with this the only factor, there is a stable equilibrium population of magnitude $K$. Other density dependent forms for the per capita birth rate, which are similar in effect if different in detail, are due to Gompertz (1825),

$$r \ln (K/H); \qquad (4.1b)$$

Smith (1963),

$$r(K - H)/(K + \epsilon H); \qquad (4.1c)$$

and Schoener (1972),

$$r[(K/H) - 1]. \qquad (4.1d)$$

The form (4.1d) is a particularly strongly stabilizing one. More general formulae which include some of the above as limiting cases are

80

$$r[(K/H)^{-g} - 1], \qquad\qquad (4.1e)$$

with $1 \geqslant g > 0$ (Rosenzweig, 1971); or

$$r[1 - (H/K)^g], \qquad\qquad (4.1f)$$

again with $1 \geqslant g > 0$ (Goel, Maitra, and Montroll, 1971). All these forms exert a stabilizing influence; all make for negative diagonal elements in the community matrix.

A quite general model, which also incorporates a falling-off of the birth rate at small populations (Allee, 1939), has been given by Watt (1960). For a review bearing on these points, see Watt (1968, Ch. 11.4).

Apart from theoretical arguments suggesting that resource limitations and other effects are likely to produce a stabilizing density dependence in the intraspecific interactions, there is a review by Tanner (1966) pointing out that, of 71 species for which adequate data were available, 46 had a negative correlation between population density and growth rate which was significant at the 95% confidence level, 15 had less significant negative correlations, 7 had no correlations (i.e. populations not significantly different from a random series), while only 1 had a significantly positive correlation—and that was the human population of the world! This work is open to the technical criticism that a negative correlation in a linear regression analysis of the logarithm of population density versus the logarithm of the previous population can be a statistical artifact (St. Amant, 1970; Maelzer, 1970). Moreover, as many of Tanner's populations were not single-species communities, his results are not always a simple measure of the diagonal elements $a_{ii}$ in the community matrix, but rather of total system stability. Even so, the general drift is in support.

(II) The rate at which predators remove prey is described by the predator's "functional response." In the Lotka-

Volterra model, this is represented by the term $-\alpha HP$ in equation (3.1a), corresponding to an unlimited attack capacity per predator, increasing linearly with prey density (i.e. as $\alpha H$). More realistic functional responses will tend to have a destabilizing influence if the predator's consumption increases less fast than linearly with increasing prey numbers, and a stabilizing influence if the response is faster than linear.

The former, destabilizing, effect is embodied for example in Ivlev's (1961) functional response,

$$\alpha HP \rightarrow k\,P(1 - e^{-cH}). \qquad (4.2a)$$

This net predation rate has the natural feature of being proportional to $H$ for small prey populations, but saturating to a constant $k$ per predator for large $H$. A very similar expression due to Holling (1965),

$$\frac{k\,H\,P}{H + D}, \qquad (4.2b)$$

has been applied in invertebrate ecology. Here $D$ refers to some given value of the prey population beyond which the predators' attack capability begins to saturate. An extreme form, which is highly destabilizing, is to let the predators each have a constant attack capacity:

$$k\,P. \qquad (4.2c)$$

A generalization of Ivlev's form, tested against field populations, is Watt's (1959)

$$k\,P[1 - \exp\,(-cHP^{1-b})]. \qquad (4.2d)$$

Another general form,

$$k\,P\,H^{g}, \qquad (4.2e)$$

with $1 \geqslant g > 0$, is given by Rosenzweig (1971).

Conversely, the predator may respond by a "switching" mechanism or by a change in its searching behavior (Hassell and Varley, 1969), in such a way as to produce a stabilizing effect. Although predator switching, as lucidly reviewed by Murdoch (1969), usually involves a model with more than two species, the effect can be built into a one-predator–one-prey model by invoking learning behavior which makes for a faster than linear functional response.

Takahashi (1964) has significantly remarked, in some detail, that the functional response may easily be of stabilizing form for small amplitude excursions from equilibrium, and destabilizing for large amplitude fluctuations. An equation of this type which is analogous to Holling's (4.2b) is

$$\frac{k\,P\,H^2}{H^2 + D^2}. \tag{4.2f}$$

Another example has been given and discussed by Watt (1959):

$$k\,P[1 - \exp\,(-cH^2\,P^{1-b})]. \tag{4.2g}$$

(III) The predator population dynamics in the Lotka-Volterra equation (3.1b) are described by a constant per capita death rate (leading to the term $-bP$), and a per capita birth rate or "numerical response" which is linearly proportional to how good life is, as measured by the prey abundance (the term $+\beta HP$). More generally, the death rate is likely to be exacerbated by relatively high predator densities, leading to stabilizing tendencies of the kind discussed under (I) above. On the other hand, a less fast than linear numerical response may well be produced by excessive prey abundance, leading to destabilizing saturation effects of the kind typified by equations (4.2a–e). Gause (1934) has suggested the predators' total numerical response may be

$$\beta HP \rightarrow \gamma PH^{1/2}. \tag{4.3a}$$

In general, the discussion of the functional response in (II) tends to apply also to the numerical response (Rosenzweig, 1971; Holling, 1959, 1961).

An interesting equation for the predator dynamics, given by Leslie (1948) and discussed by Leslie and Gower (1960) and Pielou (1969, pp. 72–74), is

$$\frac{dP(t)}{dt} = sP(t) \left[ 1 - \frac{P(t)}{\gamma H(t)} \right]. \tag{4.3b}$$

Here the growth of the predator population is of logistic form, but the conventional "$K$" which measures their resources is "$K$" $= \gamma H$, proportional to prey abundance.

(IV) In nature, the various responses and interactions rarely happen immediately, but have time delays. This destabilizing influence will be discussed later.

The above formulae constitute the basic units in a "build-a-model" toy. They may be assembled in various combinations to give one-predator–one-prey models considerably more sensible than the simple Lotka-Volterra one. Among the many possible models, two specific examples are:

$$dH/dt = rH(1 - H/K) - kP[1 - \exp(-cH)]$$
$$dP/dt = P\{-b + \beta[1 - \exp(-cH)]\}, \tag{4.4}$$

which is constructed from (4.1a), and (4.2a) for both functional and numerical responses; and

$$dH/dt = rH \left[ 1 - \frac{H}{K} \right] - \frac{kPH}{H + D}$$
$$dP/dt = sP[1 - P/(\gamma H)], \tag{4.5}$$

built from (4.1a), (4.2b), and (4.3b). Many other similar models can obviously be constructed.

## A LIMIT CYCLE THEOREM

We now proceed to a full nonlinear stability analysis of these model one-predator–one-prey communities.

Because the differential equations are nonlinear, the equilibrium or steadily maintained populations need not necessarily be constant (a stable point, the equilibrium of a marble in the bottom of a cup) as they must be for a linear system, but can alternatively be stable limit cycles wherein the population numbers undergo well-defined cyclic changes in time. The amplitude of such a limit cycle, that is the maximum and minimum values the individual populations reach during the cycle, is fixed solely by the intrinsic parameters of the model such as birth rates, predation rates, etc. So is the period, the time to complete one cycle. For a stable limit cycle, just as for a stable point equilibrium, the system if disturbed will tend to return to the equilibrium configuration. This is illustrated in Figure 4.1. Instances of stable limit cycles abound in the physical sciences, from variable stars with their pulsating luminosity to the ubiquitous Van der Pol (1928) oscillator, which has been suggested as a model for the heart; Goodwin (1970) has reviewed some examples in other areas of theoretical biology.

In essentially all models built from the above components, there is a tension between a stabilizing resource limitation or other density-dependent term (I), and destabilizing predator functional and numerical response terms (II and III). Many conventional analyses, either by analytic (Bulgakova, 1968a; Canale, 1970; Rosenzweig, 1971) or graphical (Rosenzweig and MacArthur, 1963; MacArthur and Connell, 1966; Vandermeer, 1973) means, first identify the possible equilibrium populations (i.e. the point where $dH/dt = dP/dt = 0$), and then give a linearized study of the outcome of this tension between

85

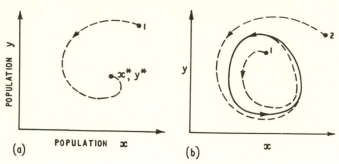

FIGURE 4.1. Depicted is the "phase space" of two species, with populations $x$ and $y$; each point in the plane corresponds to some particular value of the two populations. In (a) the point $x^*$, $y^*$ is a stable equilibrium point; if the populations are displaced from it (e.g. to point 1), they tend in time to return, as exemplified by the dashed line. In (b), the solid curve is a stable limit cycle, and in equilibrium the two populations cycle around and around this trajectory, exhibiting well-defined and periodic oscillations in the population numbers; if displaced, either inside (e.g. to point 1) or outside (e.g. to point 2) their stable limit cycle, they tend to return to it, as illustrated by the dashed lines.

stabilizing and destabilizing influences in the neighborhood of the equilibrium point. If a stable equilibrium point is not found, the model is sometimes dismissed as "unstable."

Kolmogorov (1936) went beyond this, to write the one-predator–one-prey equations as

$$dH/dt = H \ F(H,P)$$
$$dP/dt = P \ G(H,P),$$

$$(4.6)$$

and then to set out, in general terms, conditions which necessarily lead to the system's having *either* a stable point *or* a stable limit cycle. This work has recently been reviewed by Scudo (1971), and by Rescigno and Richardson (1967), who give a lucid account of the geometrical properties the isoclines $F = 0$ and $G = 0$ need to possess if they are to satisfy Kolmogorov's theorem. None of these authors gives any specific examples. MacArthur (1971a), Gilpin (1972),

Hubble (private communication), and others have found stable limit cycles to emerge from numerical studies of particular models, and Kilmer (1972) in a recent review of the subject gives an explicit analytic example of a stable limit cycle in a predator-prey system.

It is significant, however, to remark that Kolmogorov's theorem is applicable to essentially all models assembled from the components catalogued above. That is, all such models possess *either* a stable equilibrium point, *or* a stable limit cycle.

This rather robust theorem strongly suggests that those natural ecosystems which seem to exhibit a persistent pattern of reasonably regular oscillations (see, e.g., the reviews in Kormondy, 1969, p. 86, or Macfadyen, 1963, Chs. 12, 16) are in fact stable limit cycles. This is altogether different from the widespread explanation of such phenomena which associates them with the oscillations in the pathological neutrally stable Lotka-Volterra system (the stability of the frictionless pendulum), where the amplitude of oscillation depends wholly on the initial conditions (on how the pendulum was set swinging).

Specifically, Kolmogorov's theorem says that predator-prey systems of the form (4.6) have either a stable equilibrium point or a stable limit cycle, provided that $F$ and $G$ are continuous functions of $H$ and $P$, with continuous first derivatives, throughout the domain $H \geqslant 0$, $P \geqslant 0$, and that

$$\text{(i)} \quad \partial F/\partial P < 0 \qquad\qquad (4.7\text{a})$$
$$\text{(ii)} \quad H(\partial F/\partial H) + P(\partial F/\partial P) < 0 \qquad\qquad (4.7\text{b})$$
$$\text{(iii)} \quad \partial G/\partial P < 0 \qquad\qquad (4.7\text{c})$$
$$\text{(iv)} \quad H(\partial G/\partial H) + P(\partial G/\partial P) > 0 \qquad\qquad (4.7\text{d})$$
$$\text{(v)} \quad F(0, 0) > 0. \qquad\qquad (4.7\text{e})$$

It is also required that there exist quantities $A$, $B$, $C$ such that

(vi)   $F(0, A) = 0$, with $A > 0$           (4.7f)

(vii)  $F(B, 0) = 0$, with $B > 0$           (4.7g)

(viii) $G(C, 0) = 0$, with $C > 0$           (4.7h)

(ix)                  $B > C$.           (4.7i)

The above statement of the theorem is in somewhat more transparent form than the original. The proof follows (see, e.g., Minorsky, 1962, Ch. 2.9) straightforwardly from the Poincaré-Bendixson theorem, one of the key theorems of nonlinear stability analysis. We shall not present the proof, because the necessary theory is covered well in Minorsky, or, with a slant to biologists, in Rosen (1970, Ch. 5), or in sketchier but more specific form in Rescigno and Richardson (1967). What has been lacking in the literature is not the derivation of the above theorem, but rather the realization that it applies to essentially all the conventional models people use. The theorem also usually holds when certain of the above conditions are equalities ($=$) rather than inequalities ($<$ or $>$). Such cases need to be dealt with on their merits, but can often be seen to be sensible limiting cases of more general predator-prey equations which do obey the above criteria. (Thus the models of equations (4.4) and (4.5) can be seen to have either stable limit cycles or stable equilibrium points, although equation (4.4) has $\partial G / \partial P = 0$ rather than $<0$, and equation (4.5) has $H(\partial G / \partial H) + P(\partial G / \partial P) = 0$ and $C = 0$.)

In more biological terms, Kolmogorov's conditions are roughly that (i) for any given population size (as measured by numbers, biomass, etc.), the per capita rate of increase of the prey species is a decreasing function of the number of predators, and similarly (iii) the rate of increase of predators decreases with their population size. For any given ratio between the two species, (ii) the rate of increase of the prey is a decreasing function of population size, while conversely (iv) that of the predators is an increasing function. It is also required that (v) when both populations are small the prey have a positive rate of increase, and that (vi) there can be a predator population size sufficiently large to stop further prey increase, even when the prey are rare. Condition (vii) requires a critical prey population size $B$, beyond which they cannot increase even in the absence of predators (a resource or other self limitation), and (viii) requires a critical prey size $C$ that stops further increase in predators, even if they be rare; unless (ix) $B > C$, the system will collapse. These biological constraints are spelled out more fully in Scudo (1971) or Rescigno and Richardson (1967).

Even when the conditions are not fulfilled for all prey and predator populations, that is for all $H, P > 0$, they can still be useful. For example, the criterion (ii), equation (4.7b), usually requires the per capita prey birth rate to be a monotonic decreasing function of increasing $H$. This condition is violated if there is an Allee effect, whereby the per capita birth rate falls off at small $H$. This makes sense biologically; models incorporating such an effect should permit the possibility of extinction. However, Kolmogorov's theorem still applies in a restricted part of the $H - P$ plane, allowing the possibility of limit cycles wherein the prey population remains big enough for its Allee effect not to operate.

As one particular example, the theorem may be applied to the system obeying equation (4.5), to show that it possesses either a stable limit cycle or a stable equilibrium point. Then a conventional neighborhood analysis after the style of Chapter 2 reveals whether the equilibrium point is stable, whereupon we have the complete global stability character of this system laid bare. The rather messy criterion separating the regions of stable points and stable limit cycles is set out in Appendix I, and we see that a combination of relatively ineffective prey self-limitation ($K$ substantially larger than $D$) and a prey population which

89

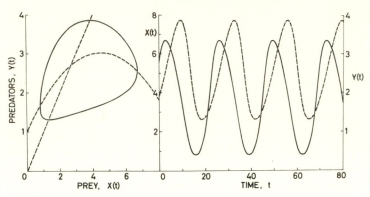

FIGURE 4.2. A specific example of a limit cycle. For the predator-prey system obeying equation (4.5), we form dimensionless variables $X(t)$ and $Y(t)$ from the prey and predator populations, respectively: $X = H/D$, $Y = P/(\gamma D)$. To the left we show the stable limit cycle endlessly traced out by these population numbers, for the parameter choice $r/s = 6$, $K/D = 10$, $K\gamma/r = 1$; the dashed lines are the isoclines of stationary prey $(dX/dt = 0)$ and predator $(dY/dt = 0)$ populations, and their intersection is the equilibrium point, here unstable. To the right we display the cyclic population numbers of prey (solid line) and predator (broken line) as functions of time, $rt$. If displaced from these trajectories, the system tends to return to them.

grows fast compared with the predator population ($r$ significantly greater than $s$) makes for stable limit cycle behavior. Figure 4.2 shows the limit cycle trajectory in the $H - P$ plane, and also the prey and predator population oscillations as functions of time, for a particular choice of model parameters. If the populations are disturbed from these stable cycles, they quickly settle back into them.

As a second example, the theorem may be similarly applied to the system (4.4). Again relatively weak prey self-regulation (relatively large $K$) leads to stable limit cycle behavior, as illustrated by Figure 5.6 (p. 135). (Generally, an increase in the environmental carrying capacity $K$ for the prey can carry the predator-prey system from a stable equilibrium point to a stable limit cycle; this transition from equilibrium point to oscillatory behaviour has been called by Rosenzweig (1971) the "paradox of enrichment.") Again the populations tend to return to this stable trajectory if perturbed from it. In the real world, random environmental fluctuations will continually supply such perturbations, and the deterministic limit cycle will represent a stable mean upon which environmental noise imposes fluctuations. This feature can be illustrated in the model (4.4) by, for example, letting the parameter $r$ have a randomly varying component. Figure 4.3 is the typical outcome of such a computation (with the white noise in $r$ having variance equal to 10% of the mean value of $r$). It is an evocative illustration.

The broad applicability of Kolmogorov's theorem has implications for those natural systems in which the populations seem to go up and down in a rather stable periodic manner. Admittedly, some of the population cycles reported in the literature are artifacts of the smoothing

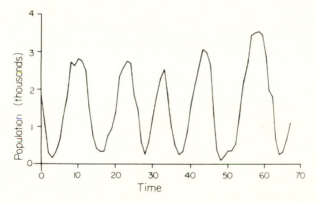

FIGURE 4.3. The prey population, as a function of time, obtained from solving the pedator-prey model (4.4) with some random environmental fluctuation. The parameters in equation (4.4) have the values $c = 10^{-3}$, $cK = 4$, $\beta = 1.5$, $k = b = 1$, and the prey intrinsic birth rate $r$ incorporates some "white noise" environmental fluctuation, having mean $r_o = 1$ and variance $\sigma^2 = 0.1$. (See also Figure 5.6 and accompanying discussion.)

procedure employed in processing the data (Cole, 1954; Slobodkin, 1961), but others are not. One may expect these genuine cycles to be stable limit cycles, sometimes with environmental fluctuations superimposed, between well-defined limits set by the interactions between and within species.

For example, Lack (1954, pp. 212–217) has suggested that the lemming population cycles in northern regions are the result of some prey-predator interaction, with the lemmings playing the role of predator, and their food the prey; we note that a lemming-vegetation system containing realistic interaction elements can naturally give rise to stable limit cycle behavior. Similarly, to regard the familiar and regular Hudson Bay lynx and hare population oscillations (Figure 4.4) as resulting from a Lotka-Volterra pure oscillation about a neutrally stable equilibrium point, which is to say having an amplitude determined by some environmental shock over 100 years ago, is absurd. This system, with the maximum hare population being constant to within a factor two over 100 years or 9 cycles, is surely the outcome of some stable limit cycle.

In general, our models suggest that predator-prey

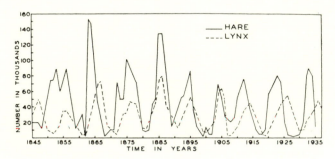

FIGURE 4.4. Changes in the populations of the lynx and the snowshoe hare, as indicated by the trading figures of the Hudson Bay Company (after Odum 1953). I reproduce this familiar figure, to play my part in embedding it in the folklore of ecology.

systems are likely to exhibit stable limit cycles when there is relatively weak intraspecific prey population regulation (relatively large $K$), and the intrinsic growth rate of the prey population exceeds that of its predators. This seems to accord with the facts reviewed by Tanner (1972) for eight pairs of prey species and their major predator. The idea that a population's dynamics may switch from a stable equilibrium point to a stable limit cycle as "life gets better," that is as effective values of $r$ or $K$ increase, is elegantly illustrated by the work of Baltensweiler (1971) on the larch bud moth in Switzerland. In their optimal habitat at altitudes between 1700 and 1900 meters these beasts have exhibited stable population oscillations since 1855, with a period of around 8 years, whereas no such phenomenon is manifest in the less favorable habitat above and below this range.

In conclusion, we mention two relevant technical points.

That systems which in a linearized stability analysis are classed as "unstable" may wind out to a stable limit cycle is easily overlooked in computer realizations of the models, because usually the ratio between the predator's minimum population and its mean population in the limit cycle is roughly of the order (May, 1972d)

$$\frac{P(\text{min})}{P(\text{mean})} \sim \exp\ [-c(K/H^*)^2]. \tag{4.8}$$

Here $c$ is a number of order unity, $H^*$ is the mean prey population, and $K$ is the maximum prey population capable of being sustained by the environmental resources: the essential assumption underlying equation (4.8) is that $K/H^*$ is large. Since computer models indeed often have $(K/H^*)^2$ substantially greater than unity, the ratio (4.8) can be so small that the predator population is below unity, and therefore extinct, before the limit cycle minimum is reached. Indeed, so long as the cycle is severe enough to carry either prey or predator numbers very low, "demographic" stochastic features of the kind discussed by Bartlett (1960) will enter, and extinction will occur sooner or later. In short, the stable limit cycle may often be of no practical relevance. This is a question that depends on the numerical details of the parameters in a particular model.

It is no accident that this discussion of the nonlinear global stability character of so wide a class of models is restricted to two-dimensional, or two-species, systems. Such a discussion is made possible by the powerful Poincaré-Bendixson theory, which unfortunately breaks down in more than two dimensions (essentially because one can no longer make an unambiguous distinction between the "inside" and the "outside" of a closed curve, as one can for curves on two-dimensional surfaces). Ecology would benefit from a mathematical breakthrough which extended this type of analysis to higher dimensions.

## TIME DELAYS

In the real world, the growth rate of a species' population will often not respond immediately to changes in its own population or that of an interacting species, but rather will do so after a time lag. Thus the effects of overgrazing may depend not so much on the contemporary herbivore population, but on an average reaching back into the past over a time roughly equal to the characteristic regeneration time for the vegetation. Alternatively, it can be that the time delay in a resource limitation effect is essentially a natality lag: one has to wait until the next generation is confronted with the havoc its predecessors wrought. Inclusion of such time delays in our model equations tends to have a destabilizing influence. This is already implicit in the discussion in Chapter 2, where we saw that a difference equation, with the enforced one-generation time lag between cause and effect, is less stable than the homologous differential equation.

The prototype model embodying this phenomenon is due to Hutchinson (1948). He considers a single species obeying a logistic growth law, with growth rate $r$ and maximum population sustainable by the environmental resources being $K$, but with a time lag $T$ built into the operation of this regulatory mechanism. We may think of this system as herbivores grazing upon vegetation, which takes time $T$ to recover:

$$\frac{dN(t)}{dt} = r\,N(t)[1 - N(t - T)/K]. \tag{4.9}$$

This equation was first studied in the economic theory of the stability of business cycles (Frisch and Holme, 1935). With no time delay, $T = 0$, of course there is the usual stable equilibrium population $N = K$. In the system (4.9) there is now a counterplay between the stabilizing density dependent resource limitation, and the destabilizing time lag. If $rT < \frac{1}{2}\pi$, there remains the stable equilibrium point at population $K$. If $rT > \frac{1}{2}\pi$, this potential equilibrium point is unstable, and there is instead a limit cycle solution, the oscillations in which become increasingly severe as $rT$ increases. (We may recall that similarly in the difference equation version of logistic growth, equation (2.26), with a generation time $\tau$, there was no stable point once $r\tau > 2$.)

The properties of the equation (4.9) have been extensively discussed in the mathematical literature. The equation may be brought into canonical form by defining

$$\tau \equiv rT, \tag{4.10}$$

and rescaling the population variable to be $x = N/K$, and the time to be $\hat{t} = rt$, to get the one-parameter family of equations

$$\frac{dx(\hat{t})}{d\hat{t}} = x(\hat{t})[1 - x(\hat{t} - \tau)]. \qquad (4.11)$$

Jones (1962a) has demonstrated the existence of limit cycle solutions in the region $\tau > 1/2\pi$. Although a complete proof that these limit cycles are stable to any perturbation is as yet lacking, Jones' (1962b) numerical studies suggest they are, and Kaplan and Yorke (1973) have used a clever extension of Poincaré-Bendixson techniques to differential-difference equations such as (4.11) which rigorously establishes stability with respect to most perturbations. The oscillation executed by the population $N(t)$ in one complete cycle is illustrated in Figure 4.5, for various values of $\tau > 1/2\pi$. Table 4.1 similarly sets out the ratios between population maximum and minimum in any one such cycle, and also the period of the cycle, as functions of $\tau$ (Martin, private communication).

Hutchinson's model has a time delay of exactly $T$ for the vegetation or whatnot to respond. More generally, and

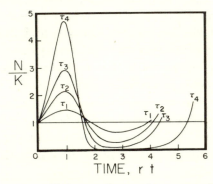

FIGURE 4.5. The oscillations undergone in one complete cycle by the population $N(t)$ whose dynamics obey the time-delayed logistic equation (4.9) with too long a time delay. For $rT = \tau < 1/2\pi$, there is a stable equilibrium point at $N/K = 1$ (the horizontal line). For $\tau > 1/2\pi$, the final periodic solutions are shown for various values of $\tau$, as indicated: $\tau_1 = 1.6$; $\tau_2 = 1.75$; $\tau_3 = 2$; $\tau_4 = 2.5$ (after Jones, 1962b).

TABLE 4.1. Properties of limit cycle solutions
of equation (4.9)

| $\tau = rT$ | $N(\max)/N(\min)$ | Cycle period |
|---|---|---|
| 1.57 | 1.00 | — |
| 1.6 | 2.56 | 4.03T |
| 1.7 | 5.76 | 4.09T |
| 1.8 | 11.6 | 4.18T |
| 1.9 | 22.2 | 4.29T |
| 2.0 | 42.3 | 4.40T |
| 2.1 | 84.1 | 4.54T |
| 2.2 | 178 | 4.71T |
| 2.3 | 408 | 4.90T |
| 2.4 | 1,040 | 5.11T |
| 2.5 | 2,930 | 5.36T |

more realistically, this time delay will depend not on the
population at some particular instant in past time, but
rather on an average over past populations; not on
$N(t - T)/K$ but rather on the weighted average

$$\int_{-\infty}^{t} N(t')Q(t - t')dt'. \tag{4.12}$$

The function $Q(t)$ specifies how much weight to attach to
the populations at various past times, in order to arrive
at their present effect on resource availability. A typical
averaging function $Q(t)$ may have the kind of shape depicted in Figure 4.6. We note that this function is characterized by some average time delay $T$, but that a spread of
times (with a width of the order of $T$ about the mode)
contributes significantly to the averaging process, that is to
the integral (4.12). Hutchinson's equation (4.9) may be recovered by taking a limiting form for the weighting function $Q(t)$, with essentially infinite height and zero width
at the time $T$; the integral (4.12) then focuses exactly on
the point $t - t' = T$, that is it singles out exactly the population $N(t - T)$.

The generalization of Hutchinson's vegetation-herbivore equation now reads

97

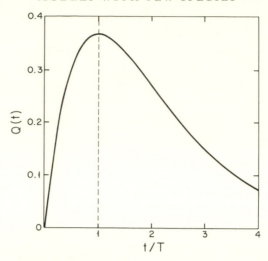

FIGURE 4.6. A typical example of a smooth time-delay averaging function $Q(t)$, as defined by the general equation (4.13), namely $Q(t) = (KT)^{-1}(t/T) \exp (-t/T)$. (In displaying $Q(t)$ as a function of $t/T$ we have omitted the constant factor $1/KT$.)

$$\frac{dN(t)}{dt} = r N(t)[1 - \int_{-\infty}^{t} N(t')Q(t - t')dt'].  \quad (4.13)$$

The stability of the possible constant equilibrium population may be studied by a neighborhood analysis, using Laplace transform techniques. A poor man's version of these methods, particularly as applied here, is in an appendix to May (1972e). It is amusing to note that the more realistic smooth time delay function of Figure 4.6 permits an easier analysis than the mess consequent upon the use of Hutchinson's original singular "time delay at exactly $T$" of equation (4.9). It is not often that greater realism makes for easier mathematics.

The conclusion remains that a system of this general type does, or does not, have a stable equilibrium point, depending on whether the characteristic growth rate time $T_1$, defined as $T_1 = 1/r$, is large or small compared with the time delay $T$ in the regulatory mechanism. That is, give or take a numerical constant of order unity, the equilibrium point is

$$T_1 > T: \text{stable} \atop T_1 < T: \text{unstable} \Bigg\} \qquad (4.14)$$

In the event that there is no stable equilibrium point, the nonlinear analysis usually leads again to a stable limit cycle (Dunkel 1968 a, b).

This conclusion is very much in accord with the precepts of engineering control theory. The instantaneous density-dependent form $(1 - N/K)$ represents a stabilizing negative feedback in the system, and it is a central result of control theory that, if such feedback occurs with a time lag longer than the natural period of the system (i.e. $T > T_1$),

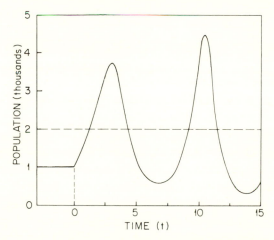

FIGURE 4.7. A computed example of the growing population oscillations which follow from the logistic equation (4.9) with too long a time lag in the density dependent mechanism: here $r = 1$ and $T = 3\pi/5$, so that $rT$ just exceeds $\frac{1}{2}\pi$. The (unstable) equilibrium population is $K = 2,000$, and in the example the population is controlled to be half this value for a while, and then let go.

99

the upshot will be instability. For a discussion, complete with graphical examples, see Astrom (1970, pp. 9 and 178–179). Figure 4.7 is a numerical solution of Hutchinson's equation (4.9) with too large a time delay, and it is reminiscent of many graphs of population "crashes" in the ecology literature. In this figure, the population is initially constrained to be constant, and then let go, whereupon a pattern of growing oscillations builds up toward a stable limit cycle, which may easily be so severe (see Table 4.1) as to produce extinction. Hutchinson (1954) was moved to conclude: "These considerations should lead us to expect oscillatory changes in single species populations even under relatively stable environmental conditions as normal events."

## NICHOLSON'S BLOWFLIES

From the discussion so far, stable limit cycles emerge as ubiquitous solutions to the simplest equations of population ecology.

As a further illustration, we apply equation (4.9) to Nicholson's (1954) classic laboratory experiments with the Australian sheep-blowfly, *Lucilia cuprina.*

An attempt to build a realistic model for these populations would need to take into account separate age classes and a host of relevant aspects of blowfly biology. However, equation (4.9) represents an extremely crude first approximation, incorporating the minimum amount of essential biological information about the system, namely: (i) the blowfly population has an intrinsic rate of increase, $r$; (ii) there is a resource limitation, $K$, set by the supply of ground liver; and (iii) this resource limitation acts with a time delay, $T$, roughly equal to the time for a larva to mature into an adult. A fuller description, in terms of specific age classes, is roughly approximated by this single equation.

100

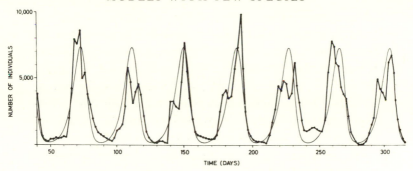

FIGURE 4.8. From the one-parameter family of limit cycles generated by the time-delayed logistic equation (4.9) (see Figure 4.5
and Table 4.1), we display that which best fits the oscillations in
Nicholson's blowfly populations. The experimental data are from
Nicholson (1954); the theoretical curve, with $rT = 2.1$, is in good
agreement considering the crudity of the model.

In fitting this model to the data, we have only the dimensionless parameter $rT$ at our disposal; $K$ is absorbed in
setting the scale of the $y$-axis for the population, and $r$ in
scaling the $x$-axis for the time (cf. equation (4.11)). This
single parameter $rT$ is completely determined by Nicholson's data (Figure 4.8) because it depends sensitively on
the ratio between maximum and minimum values of the
oscillating population (cf. Table 4.1). We estimate $rT$
$\sim 2.1$. Thence, again consulting Table 4.1, the appropriate theoretical period may be compared with the observed experimental oscillation period, to conclude that
the time delay $T$ is roughly 9 days. In fact, these blowfly larvae take around 11 days to become adult (Nicholson,
1957, Figure 6). The theory also predicts that if the amount
of ground liver be doubled, that is if $K$ be doubled, then
everything should be exactly as before, except that the
scale of the $y$-axis should be doubled. This is precisely what
Nicholson (1954, p. 22) found.

Figure 4.8 shows the comparison between Nicholson's
data and the curve obtained from equation (4.9). In view

of the crudity of the model, the agreement is surprisingly good. It suggests that stable limit cycles generated by the time-delayed regulatory mechanism are indeed the basic feature in the dynamics of these populations. A similar, but somewhat more realistic, model with two free parameters has been discussed in connection with Nicholson's experiments by Maynard Smith (1968, pp. 52–55); no explicit fit to the data was attempted.

We conclude by speculating that there is a natural mechanism whereby some simple populations may come to have values of $rT$ slightly in excess of that value ($1/2\pi$ for equation (4.9)) which initiates stable limit cycle behavior. So long as there remains a stable equilibrium point, natural selection may tend to increase $r$. Thus $rT$ climbs a hill ($rT$ increases towards $1/2\pi$) until it reaches the ridge separating stable equilibrium point from stable limit cycle behavior ($rT = 1/2\pi$), and thereafter increasing $r$ precipitates the system into increasingly severe limit cycle oscillations. We may expect $rT$ to settle down at some value just past the critical one, before the oscillations become too severe. Although this plausibility argument makes implicit appeal to group selection, such an extreme limit cycle situation, with populations periodically being carried to very low values, is an ideal situation for models of group selection to operate successfully (Levins, 1970a; Boorman and Levitt, 1972).

## TIME DELAYS AND THE VEGETATION-HERBIVORE-CARNIVORE SYSTEM

We end this chapter by wedding together the above models for predator-prey with no time delays in the interactions, and for single species with time-delayed resource limitation. The latter models were visualized as an herbivore population regulated by a vegetation resource.

102

The addition now of a predator trophic level gives models which may contain some of the basic features of a vegetation-herbivore-carnivore system.

Since our interest is the qualitative effects associated with the addition of another trophic level onto our earlier vegetation-herbivore system, we describe the predator-prey interactions by the simplest model containing their essentials, namely the Lotka-Volterra one. Then with Hutchinson's original form (4.9) for the resource limitation effect, the equations for the herbivore and carnivore populations, $H(t)$ and $P(t)$ are

$$\left.\begin{aligned}
\frac{dH(t)}{dt} &= r\,H(t)[1 - H(t-T)/K] - \alpha\,H(t)\,P(t) \\[2ex]
\frac{dP(t)}{dt} &= -b\,P(t) + \beta\,P(t)\,H(t).
\end{aligned}\right\} \quad (4.15)$$

Similar equations follow for the more general "smoothed out" time delay of equation (4.12). By neglecting the resource limitation effect, $K \to \infty$, we recover the Lotka-Volterra system (3.1), with its purely oscillatory stability character, as illustrated by Figure 3.2.

Greater realism can be achieved by replacing (4.15) with equations which embody predator functional and numerical responses along the lines discussed at the beginning of the chapter. The stability character of such more realistic equations has been discussed in some detail (May, 1972e), and the general conclusions are similar to those based on the simple model (4.15).

In equation (4.15), and also in the more realistic models discussed by May, the predator-prey interactions ($\alpha HP$ and $\beta HP$ in equation (4.15)) do not themselves incorporate any time delay. This corresponds not so much to setting any such delay to be zero, but rather to assuming it to be small compared to all other time scales in the problem (which in this instance will be seen to be only $T$ and $(r\,b)^{-1/2}$).

There would seem to be evidence that at least in some circumstances the predator's functional and numerical responses do have such short time lags. Thus in their extensive study of freshwater animal communities, Hall, Cooper, and Werner (1970) concluded that the dominant predator, the bluefish, had a quick response to numerical changes in the prey; this bluefish plays a role closely analogous to that of *Pisaster* in Paine's (1966) studies. The inclusion of time delays in the predator-prey interactions can be carried out, at least to a good approximation, and it has qualitatively the same effect as the destabilizing predator functional response treated by May. In this context it is appropriate to refer to the work of Wangersky and Cunningham (1957), which would seem to be the only previous work on three-level systems incorporating time lags. These authors deal with a vegetation-herbivore-carnivore system where the only time delays are in the prey-predator interaction term. Their work thus has the unusual feature that the prey-predator systems are invariably less stable than those with prey alone. Were this a feature of the real world, resource management would be easier. Goel, Maitra, and Montroll (1971) have recently revisited Wangersky's and Cunningham's model, and corrected some faults in the mathematics, but as the basic premise seems ill-chosen this is somewhat beside the point.

We further assume that, as often happens in the real world, the equilibrium prey population in the total predator-prey system (4.15), $H^* = b/\beta$, is significantly less than that set by environmental resources alone, $K$. The natural time scale in this system is then essentially

$$T_2 = (r\,b)^{-1/2}, \tag{4.16}$$

which is to say the geometric mean of the intrinsic time scales for herbivore and for carnivore populations. This time scale is recognizable as the Lotka-Volterra oscillatory period. More generally, if there is a destabilizing saturation of the predators' appetite or some like effect to produce divergent oscillations in the isolated prey-predator system, $T_2$ still characterizes the time between successive population peaks.

The vegetation-herbivore-carnivore system modeled by (4.15), or by similar but more realistic equations, may be shown to have an equilibrium point which is stable or unstable depending on whether the time delay $T$ is small or large compared to this characteristic time $T_2$ (May, 1972e). That is, again give or take a numerical constant of order unity, the equilibrium point is

$$T_2 > T: \text{stable}$$
$$\qquad\qquad\qquad\qquad (4.17)$$
$$T_2 < T: \text{unstable.}$$

Figure 4.9 illustrates the effect of introducing into a predator-prey model, of the oscillatory kind depicted in Figure 3.2, the resource limitation term which, in the absence of a predator, gave Figure 4.7. The total system (which is chosen to have $T_2 > T$) is quite stable, with the herbivore population returning to its equilibrium value after being disturbed.

Putting together the result (4.14) without predators and the result (4.17) with predators, we arrive at a significant conclusion. Suppose we have a community comprising vegetation-herbivore-carnivore, with the natural time scales: $T$ for the typical vegetation recovery time after grazing (the time delay in the resource limitation effect); $T_1$ for the herbivore birth rate time scale; $T_2$ for the geometric mean of herbivore birth rate and carnivore death rate times, that is for the characteristic oscillatory period of the predator-prey system in the complete absence of any stabilizing density dependent terms. Then if roughly

$$T_1 < T < T_2 \qquad\qquad (4.18)$$

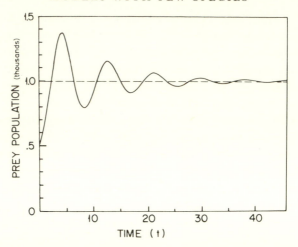

FIGURE 4.9. This figure illustrates the stable herbivore population behavior which can occur in a vegetation-herbivore-carnivore model built up from unstable components of the kind illustrated in Figures 3.2 and 4.7. That is, it shows a numerical solution of equation (4.15) with $r = 1$, $b = 0.2$, $T = 3\pi/5$, and $K\beta/b = 2$; as in Figure 3.2 the prey population is plotted as a ratio to the equilibrium population $H^*$ (thus the dashed line represents this equilibrium population), and the initial perturbation is again $H(0)/H^* = 0.5$, $P(0)/P^* = 1.0$. Essentially, the system is now stable because, in contrast with Figure 4.7, the natural time scale is longer than the time lag (cf. equations (4.16) and (4.17)).

the total *vegetation-herbivore-carnivore system is stable, whereas the vegetation-herbivore system with no predators is unstable.* In the former case the delay time is relatively short, and so the feedback is stabilizing. In the latter case the delay time is relatively long, and the feedback is destabilizing.

At least for mammalian predator-prey systems, and possibly for a wider class, it seems that $T_1$ is often less than or of the order of $T$, whereas the predator death rate ($b$) is substantially smaller than the prey birth rate ($r$) so that $T_2$ is appreciably larger than $T_1$. Hence in such real biological communities we expect—and often find—that the total vegetation-herbivore-carnivore system is stable, while the simple vegetation-herbivore system, with predators removed, is not. (The Isle Royale unstable vegetation-moose and subsequently stable(?) vegetation-moose-wolf numerical data (Mech, 1966; Dixon and Cornwall, 1970) bear out these suggestions in detail in the way that theoreticians are fond of calling typical.)

Furthermore, this model hints at a reconciliation between some of the seemingly opposed views expressed in the celebrated controversy between Hairston, Smith, and Slobodkin (1960, 1967) and their critics (Murdoch, 1966; Ehrlich and Birch, 1967). We notice that although the equilibrium herbivore population is set by the predators and can be much smaller than the maximum capable of being sustained by the vegetation, so that "the world is green," nonetheless it is the resource limitation term "$N/K$" which makes the equilibrium point a stable one. The answer to the question whether it is available vegetation or predators that "control" the herbivores may be that both do: if the condition (4.18) prevails, the vegetation-herbivore system alone is unstable (as in Figure 4.7), the herbivore-carnivore system alone is unstable or at best purely oscillatory (as in Figure 3.2), while the overall vegetation-herbivore-carnivore system is quite stable (Figure 4.9).

In summary, the basic point is that if a biological system has a potentially stabilizing negative feedback mechanism, which is applied with a time delay long compared to the natural time scale of the system, the result will be instability, not stability. However, by adding an extra trophic level we may alter the system's characteristic time scale, making it longer than the delay time, and so taking advantage of the feedback stabilization. This is a common theme in systems control theory, and it provides an interesting example of a recondite strategy whereby increased trophic complexity can enhance stability.

## SUMMARY

Stable limit cycle behavior constitutes a natural explanation for the population cycles which are sometimes found in nature. Such stable limit cycles can be shown to be implicit in essentially all moderately realistic predator-prey models.

Limit cycles can also arise in a system where a potentially stabilizing negative feedback is applied with a time lag long compared to the natural time scale of the system. Nicholson's blowfly experiments are discussed in this light.

This theme is developed further for vegetation-herbivore and vegetation-herbivore-carnivore systems in which the vegetation's stabilizing resource limitation effect operates on the herbivore population with a time delay. Under certain conditions, which are commonly met in nature, the vegetation-herbivore system with no predators has no stable point, but the vegetation-herbivore-carnivore system does have a stable point. This model suggests a specific mechanism whereby herbivore population numbers may often be set neither by predators alone nor by vegetation alone, but by an explicit interplay between both effects.

# CHAPTER FIVE

# Randomly Fluctuating Environments

So far, all the models have assumed an unvarying, deterministic environment. But real environments are uncertain, stochastic. The birth rates, carrying capacities, competition coefficients, and other parameters which characterize natural biological systems all, to a greater or lesser degree, exhibit random fluctuations. Consequently equilibrium is not the constancy of the physicist, but rather an average around which the system fluctuates. Elton (1958) observes that the "chief cause of fluctuations in animal numbers is the instability of the environment. The climate in most countries is always varying. . . ." A trenchant affirmation of the view that such environmental vagaries need be incorporated is Ehrlich's and Birch's (1967) "models must be stochastic not deterministic." For fuller discussion along these lines, see Macfadyen (1963, p. 181), Kormondy (1969, p. 84) or Levin (1970).

For deterministic environments, we seek the constant equilibrium populations $N_i^*$, and then study their stability, which follows from the dynamics of the interactions between and within species. In particular, for relatively small amplitude disturbances, the interactions are summarized and the stability set by the community matrix $A$, as discussed in Chapter 2.

Once environmental stochasticity is admitted, so that some of the parameters are fluctuating randomly about their mean values, we obviously can no longer speak of *the* population $N(t)$ at time $t$, but only of its probability distribution. Such a distribution function, $f(n, t)$, gives the probability to observe $n = 0$, $1$, $2$, $\ldots$, $N$, $\ldots$ animals at time $t$. More generally, for a community with $m$ species, there will be a multivariate probability distribution function, $f(n_1, n_2, \ldots, n_m; t)$. By taking the usual statistical moments of this distribution, we get the mean numbers of animals at $t$; these may, or may not, be equal to the deterministic populations $N_i(t)$ obtained from the deterministic equations in which all the environmental parameters are fixed at their mean values. Similarly one may obtain the variances of the fluctuating populations, and so on.

Furthermore, the analogue of *the* equilibrium populations, $N_i^*$, which were the time-independent solutions of the deterministic population equations, is the time-independent probability distribution function, $f^*(n)$. *This equilibrium probability distribution is to the stochastic environment as the stable equilibrium point is to the deterministic one.*

In the deterministic case, we studied the response to a specific disturbance from the equilibrium configuration; in the stochastic environment, an incessant sequence of such perturbations is built into the fabric of the model. The relation between deterministic and stochastic cases is illustrated for two interacting populations, $N_1$ and $N_2$, in Figure 5.1. In the deterministic environment, Figure 5.1(a), we show a stable equilibrium point, corresponding to constant populations $N_1^*$ and $N_2^*$. If disturbed from this point, the dynamics of the populations' interaction brings the system back: for small disturbances, these dynamics are described by the community matrix, and the characteristic time to return to equilibrium is measured by the (negative) real parts of the matrix's eigenvalues $\lambda$. In the stochastic environment, Figure 5.1(b), there is no longer an equilibrium point, but rather a probabilistic "smoke cloud," described by the equilibrium probability distribution. There is now a

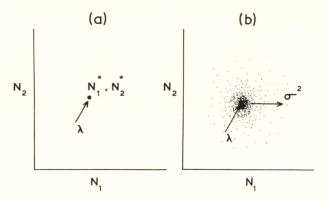

FIGURE 5.1. Schematic representation of the character of an equilibrium two-species community in (a) a deterministic and (b) a stochastic environment. In (a) we have a stable equilibrium point, corresponding to populations $N_1^*$ and $N_2^*$: the community dynamics, as measured by the eigenvalues $\lambda$ of the community matrix, return the system to its equilibrium point if it is perturbed. For the corresponding stochastic environment of (b), the equilibrium community is described by some time-independent probability distribution: this probability cloud is in tension between the stabilizing influences of the interaction dynamics (again measured by the eigenvalues $\lambda$), which act to compact the cloud, and the destabilizing environmental fluctuations (measured by $\sigma^2$), which act to disperse the cloud.

continuous spectrum of disturbances, generated by the environmental stochasticity, and the system is in tension between two countervailing tendencies. On the one hand, the random environmental fluctuations (measured by a characteristic variance $\sigma^2$) act to spread the cloud, to make the probability distribution more diffuse, while on the other hand the dynamics of the stabilizing population interactions tend to restore the populations to their mean values, to compact the cloud. For a relatively compact probability cloud, the interaction dynamics are again measured by the eigenvalues of the same community matrix as for the deterministic case, the matrix being evaluated using the mean values of the environmental parameters.

We see immediately that, if the stabilizing effects of the interactions are "strong" compared to the diffusive effects of the random environmental fluctuations, the probability cloud will be compact, and may for many practical purposes be indistinguishable from the deterministic equilibrium point. In such cases, the deterministic model will be entirely relevant. Conversely, if the stability provided by the population interactions is "weak" compared with the environmental variance, then (even though the deterministic model may be eminently stable) the probability cloud is highly dispersed, with a significant likelihood that one or both species' population may vanish. In such a case, the deterministic model clearly is irrelevant.

This point can be developed further with another impressionistic picture. For models with deterministic environmental parameters, we have a stable neighborhood of the equilibrium point if and only if all eigenvalues of the community matrix lie in the left-hand half of the complex plane: Figure 5.2(a). For the corresponding model with stochastic parameters, this condition is necessary, but insufficient, for the existence of a relatively compact equilibrium probability cloud for the population. It is required, in addition, that the stability provided by the interactions, which is measured by the real parts of the community matrix eigenvalues, be sufficiently great to counteract the diffusive effects of the random fluctuations. Thus the eigenvalues must all lie to the left of the imaginary axis by an amount measured roughly by the degree of environmental variance: Figure 5.2(b). In the deterministic environment, we require only that the terrain slope everywhere upward from a stable equilibrium valley (which is ensured by Figure 5.2(a)). In the stochastic environment, the landscape is heaving up and down like the floor of a fun-house, and the intrinsic upward slope must be great enough to forbid any significant fluctuation to a downward slope (which is ensured by Figure 5.2(b)).

In deterministic circumstances, "stability" usually refers to the propensity to return to an equilibrium point. In stochastic circumstances, it intuitively seems appropriate to refer to systems characterized by large fluctuations in the population numbers as "unstable," and to those with relatively small fluctuations as "stable." These usages may be related by recalling the definition of the quantity $\Lambda$ (equation (2.18)), which measures how far to the left of the imaginary axis the largest eigenvalue of the community matrix lies. Neighborhood stability in the mechanical, deterministic sense in an unvarying environment simply requires $\Lambda > 0$. In a stochastic environment, whose random fluc-

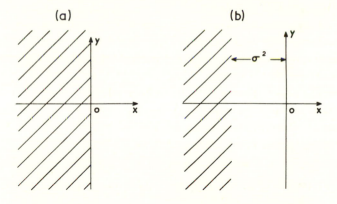

FIGURE 5.2. The eigenvalues $\lambda$ of the community matrix $A$ may be represented as points $(x, y)$ in the complex plane. In the deterministic case, (a), the criterion for an equilibrium community to be stable with respect to small disturbances is that all such eigenvalues have negative real parts, i.e. lie in the hatched area of (a). For a stochastic environment, (b), with a variance characterized by $\sigma^2$, it is necessary for the eigenvalues to lie far enough into the left-half plane for their stabilization to countervail against the diffusive effects of the random fluctuations, i.e. they need to lie in the hatched area of (b).

tuations are typified by some variance $\sigma^2$, the stability provided by the population interaction dynamics is again characterized by $\Lambda$. It is no longer enough that $\Lambda > 0$, for if $\Lambda$ is significantly less than $\sigma^2$, the populations will exhibit large fluctuations, rapidly leading to (local) extinction. In the intermediate region where $\Lambda$ and $\sigma^2$ are commensurate, the populations are likely to undergo significant fluctuations, even though they persist for long times. Finally, if $\Lambda$ is much greater than $\sigma^2$, the population fluctuations are relatively small, and the environment is effectively determinate. Thus the rather intuitive usage, which measures "stability" or "instability" by the relative magnitude of the population fluctuations in a stochastic environment, is tied to the ratio between $\Lambda$ and $\sigma^2$. It depends on the balance of power between the countervailing forces of stabilizing population interactions and randomizing environmental fluctuations.

The remainder of the chapter substantiates the above assertions. First, a general equation for the probability distribution function is given, and then examples of environmentally stochastic systems with 1, 2, and $m$ species are discussed. The presentation here is self-contained only with respect to the broad conclusions: fuller accounts of the details of the various examples are in Lewontin and Cohen (1969), Levins (1969a, b), and May (1972f).

## THE FOKKER-PLANCK OR DIFFUSION EQUATION

Consider the single species population whose dynamics are described by equation (2.1),

$$\frac{dN(t)}{dt} = F(N(t)). \tag{5.1}$$

If now the environment is randomly fluctuating, so that one or more of the parameters in this general equation are stochastic variables, we need to reformulate the mathematics in terms of the probability distribution function $f(n, t)$. In the interesting and general case when the variability is "white noise" (an assumption discussed below), the partial differential equation for the probability distribution is called, depending on the author's background, the Fokker-Planck, or the Kolmogorov, or simply the diffusion equation:

$$\frac{\partial f(n, t)}{\partial t} = -\frac{\partial}{\partial n}(M(n) f(n, t)) + \frac{1}{2}\frac{\partial^2}{\partial n^2}(V(n) f(n, t)). \quad (5.2)$$

Here $M(n)$ is defined as the mean value, and $V(n)$ as the variance, of the right-hand side of equation (5.1):

$$M(n) \equiv \langle F(n) \rangle \quad (5.3)$$

$$V(n) \equiv \langle (F(n) - M)^2 \rangle. \quad (5.4)$$

These averages are taken over all the stochastic parameters in the function $F$. The equilibrium probability distribution, if it exists, is then given by

$$0 = -\frac{d}{dn}(M(n) f^*(n)) + \frac{1}{2}\frac{d^2}{dn^2}(V(n) f^*(n)). \quad (5.5)$$

More generally, if the $m$-species community of equations (2.8),

$$\frac{dN_i(t)}{dt} = F_i(N_1(t), N_2(t), \ldots, N_m(t)), \quad (5.6)$$

incorporates some white noise environmental fluctuations, the joint probability distribution $f(n_1, n_2, \ldots, n_m; t)$ will obey the appropriate generalization of equation (5.2):

$$\frac{\partial f}{\partial t} = -\sum_{i=1}^{m}\frac{\partial}{\partial n_i}(M_i f) + \frac{1}{2}\sum_{i,j=1}^{m}\frac{\partial^2}{\partial n_i \partial n_j}(V_{ij} f). \quad (5.7)$$

Here, in logical extension of the above definitions,

115

$$M_i(n_1, n_2, \ldots, n_m) \equiv \langle F_i \rangle \qquad (5.8)$$

$$V_{ij}(n_1, n_2, \ldots, n_m) \equiv \langle (F_i - M_i)(F_j - M_j) \rangle. \qquad (5.9)$$

These averages are taken with respect to all fluctuating environmental parameters. Again the equilibrium probability distribution $f^*(n_1, n_2, \ldots, n_m)$, if it exists, is the solution of equation (5.7) with the left-hand side equal to zero.

In equations (5.2), (5.5), and (5.7), the first term on the right-hand side is the "drift" or "friction" term. It represents the interaction dynamics of the system, and, if the deterministic system is stable, this term seeks to prevent the probability cloud from fuzzing out; it corresponds to the inward arrow in Figure 5.1(b). Conversely, the second term on the right-hand side in (5.2), (5.5), and (5.7) is the "diffusion" term, which derives from the environmental stochasticity, and acts to disperse the cloud; it corresponds to the outward arrow in Figure 5.1(b).

In the limiting case of *no* environmental variability, it can be shown that equations (5.2) and (5.7) collapse back to (5.1) and (5.6), respectively, and that the cloud of Figure 5.1(b) shrinks to the point of Figure 5.1(a).

The Fokker-Planck equation rests on the assumption that the random fluctuations in the environmental or other parameters are "white noise." That is, the random distribution from which the fluctuations are drawn is the same at all times (no systematic evolutionary change in the environment), with a variance measured by $\sigma^2$, say, and there is no correlation between the fluctuations at successive instants. The assumption of "white noise" corresponds to specific assumptions about the higher moments of the noise spectrum, which in turn lead to the Fokker-Planck equation.

In practice, this "no serial autocorrelation" assumption means only that fluctuations be correlated over times short compared to all other relevant time scales in the system.

An equivalent alternative statement is that if the noise spectrum be resolved into its constituent frequency components, "white noise" is that with all frequency components equal in magnitude (hence the terminology: there is no preferred frequency, no "color"). In a sense, white noise, with its absence of temporal correlations, is the opposite extreme from the deterministic case, where the temporal correlations are perfect. We may thus hope to "bracket" reality between these extremes. Interesting complications can arise for a nonwhite noise spectrum with a correlation time of the same order as some natural time scale in the system; that is for a noise spectrum with a "color" which resonates with some natural frequency in the system. This interesting question is rarely raised in the literature on stochastic equations, which usually assumes white noise without further discussion, and we shall follow suit by not pursuing these complications here.

Appendix IV contains some more technical discussion of these points.

## SINGLE SPECIES MODELS WITH NO STEADY STATE

The simplest model containing environmental randomness is the pure exponential growth process with a fluctuating growth rate, as exhibited in equation (2.37). The analogous difference equation, for a population of discrete generations growing exponentially at a fluctuating rate, has been studied by Lewontin and Cohen (1969: equation (2.22) with $r$ a random variable). They press the counterintuitive point that "If a population is growing in a randomly varying environment, such that the finite rate of increase per generation is a random variable with no serial autocorrelation, . . . even though the expectation of population size may grow indefinitely large with time, the probability of extinction may approach unity." A precisely similar conclusion holds for the continuous growth model of equation (2.37) (May, 1973). Lewontin and Cohen's remarks have recently been corroborated by numerical simulation studies (Roff, 1972).

The deterministic exponential growth process does not admit of an equilibrium point (except for the knife-edge case of zero growth rate), and is consequently less interesting than those which do. Even so, it is worth noting the basic message emerging from these simplest models of population growth in a randomly varying environment, namely that such fluctuations do not merely enhance the possibility of extinction (which is intuitively obvious), but, if large enough, will produce extinction even in a population whose expectation value is increasing exponentially.

## SINGLE-SPECIES MODELS WITH A STEADY STATE

The paradigm for a single-species model with a stable equilibrium point is the familiar logistic equation of population growth (cf. equation (2.7)),

$$\frac{dN(t)}{dt} = N(t) \, [k - N(t)]. \tag{5.10}$$

Here $k$ represents the carrying capacity, and the time has been rescaled to absorb the conventional factor $r/k$ (assumed to be positive). Levins (1969b) has catalogued the separate effects of introducing stochasticity into the factors $r$, $k$, and a combination of the two; his results are similar to, but less explicit than, ours. We choose to work with the form (5.10) because it is mathematically transparent, and is in the form which arises naturally in Chapter 6.

In the deterministic case, the community matrix in the vicinity of the point $N^* = k$ has the single eigenvalue $-k$, which is to say $\Lambda = k$; stability follows so long as $\Lambda > 0$ (see p. 20).

Now suppose the environmental parameter $k$ varies randomly,

$$k = k_0 + \gamma(t). \tag{5.11}$$

Here $k_0$ is a constant, being the mean value of $k$, and $\gamma(t)$ is white noise with mean zero and variance $\sigma^2$. The appropriate Fokker-Planck equation for the population probability distribution $f(n, t)$ may be written down, applying the recipes (5.3) and (5.4) to calculate the "friction" and "diffusion" coefficients corresponding to equation (5.10) with (5.11). The steady equilibrium probability distribution $f^*(n)$, if it can sensibly be said to exist, is then obtained by putting the left-hand side of the Fokker-Planck equation equal to zero, as in the general equation (5.5). The resulting equilibrium probability function is a standard "Pearson Type III" gamma distribution

$$f^*(n) = C[n]^{2(k_0/\sigma^2)-2}e^{-(2n/\sigma^2)}, \tag{5.12}$$

provided

$$k_0 > \tfrac{1}{2}\sigma^2. \tag{5.13}$$

There is no equilibrium solution otherwise. $C$ is the normalization constant, making the integrated probability unity. The derivation of this result is sketched in Appendix V.

This probability distribution is illustrated in Figure 5.3. The area under this curve between $n = 0$ and $n = 1$ represents the probability that, in the real world where animal numbers come with integer values, there will be extinction; clearly if the average carrying capacity $k_0$ is a large number this extinction probability can be negligibly small, provided $k_0$ significantly exceeds $\tfrac{1}{2}\sigma^2$.

From this probability distribution it follows that the

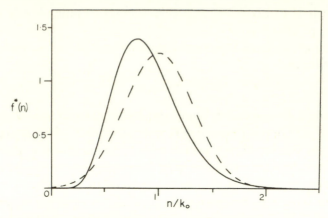

FIGURE 5.3. The equilibrium probability distribution, $f^*(n)$, for a single-species population, $n$, whose dynamics are described by the differential equation (5.10) with the stochastic parameter (5.11). The population is displayed as a ratio to the deterministic equilibrium population, i.e. as $n/k_o$. The environmental variance here is $\sigma^2 = (0.2)k_o$. The illustration bears out the discussion in the text, particularly the displacement of the mean $(0.9\ k_o)$ and modal $(0.8\ k_o)$ populations to values below the deterministic one $(k_o)$.

The dashed line depicts the corresponding approximate probability distribution of equation (5.17). The approximation is seen to be reasonable, even for this substantial (20%) level of environmental variance.

mean population in the community described by equation (5.10) in a stochastic environment is

$$\langle n \rangle = k_0[1 - (\sigma^2/2k_o)]. \tag{5.14}$$

The root-mean-square relative fluctuation of the population about this mean can likewise be shown to be

$$\frac{\sqrt{\langle (n - \langle n \rangle)^2 \rangle}}{\langle n \rangle} = \left\{ \frac{(\sigma^2/2k_o)}{1 - (\sigma^2/2k_o)} \right\}^{1/2}. \tag{5.15}$$

We note that the relative fluctuations become increasingly severe as $\sigma^2$ increases towards $2k_o$, beyond which no equilibrium solution exists.

Further illustration is provided by Figure 5.4, which

shows the fluctuating population obtained from a direct numerical solution of equation (5.10) itself, using a random number generator to give the stochastic parameter $k$ of equation (5.11).

These exact nonlinear results all bear out the general remarks made at the beginning of the chapter. The $1 \times 1$ community matrix in this stochastic case, evaluated using the average environmental parameters, has the eigenvalue $-k_0$. Consequently the familiar quantity $\Lambda$, which measures the stabilizing effects of the population interactions, is just $\Lambda = k_0$. We see from equation (5.13) that, for $\Lambda < \frac{1}{2}\sigma^2$, the environmental fluctuations are so significant that no steady probability distribution exists: the system is doomed to speedy extinction. For $\Lambda > \frac{1}{2}\sigma^2$, but not

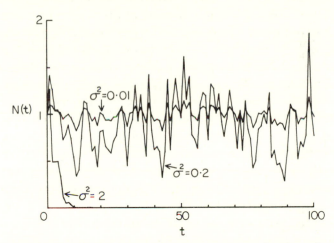

FIGURE 5.4. The fluctuating population $N(t)$ as a function of $t$, obtained directly from computation on equation (5.10). We repeat the same run of random numbers for $k$, characterized however by three different variances. Rescaling the deterministic equilibrium population to be unity ($k_0 = 1$), we see that for $\sigma^2 = 0.01$ fluctuations are small, for $\sigma^2 = 0.2$ they are appreciable (this being the case illustrated in Figure 5.3), while for $\sigma^2 = 2$ the population is soon extinguished (the criterion (5.13) is violated, and there is no equilibrium probability distribution).

much greater, the fluctuations about the long term average population are severe. As $\Lambda$ becomes substantially greater than $1/2\sigma^2$, the fluctuations become relatively unimportant, and in the limit $\Lambda \gg 1/2\sigma^2$ the deterministic solution is recovered. This is as suggested by Figure 5.2(b).

Another consequence of our nonlinear analysis is the observation that fluctuations in the carrying capacity result in a long term average population number which is less by a factor $[1 - (\sigma^2/2k_o)]$ than would be deduced from simply using the mean carrying capacity in a deterministic population equation. Thus, increasing environmental fluctuations hurt a population in two ways, as they both increase the relative severity of population fluctuations and depress the average population numbers. This feature of a stochastic environment has been pointed out by Levins (1969a, 1970a). Although arriving at his conclusions from a quite different viewpoint, and using very different analytic techniques, Thomas (1971) has similarly observed that communities inhabiting fluctuating environments will tend to maintain an average population below what he calls the "malthusian" value, which roughly corresponds to the deterministic mean; Thomas adduces some data from primitive human societies to support this view. The extensive numerical simulations carried out by Roff (1972) for the extinction and persistence of populations with density-dependent random growth rates also substantiate the tendencies remarked above.

Finally, although an exact solution of the nonlinear stochastic equation was possible in this example, it is interesting to look at an approximate solution, which has validity when the probability cloud is not too diffuse.

Let the quantity $\nu$ measure the fractional fluctuation of the population from its deterministic value, $N^* = k_o$:

$$\nu = \frac{n - N^*}{N^*} = \frac{n - k_o}{k_o}. \tag{5.16}$$

If, in the formulae (5.3) and (5.4) for the friction and diffusion coefficients, we now use a linearized approximation for $F(n)$, in which terms of order $\nu^2$ and higher are discarded, we obtain a Fokker-Planck equation for the approximate equilibrium probability distribution $\hat{f}*(\nu)$. The equation, which represents a valid approximation so long as the equilibrium probability cloud is not too diffuse, gives rise to the gaussian distribution (Appendix V)

$$\hat{f}*(\nu) = [k_0/\pi\sigma^2]^{1/2} \exp(-k_0\nu^2/\sigma^2). \qquad (5.17)$$

That is, the fluctuating population is approximately distributed normally, with mean $\langle \nu \rangle = 0$ or $\langle n \rangle = k_0$, and with variance $\sigma^2/2k_0$. This approximate distribution is displayed along with the fully exact solution in Figure 5.3.

Notice that this approximation gives an excellent qualitative description of the exact results discussed above. On the basis of the approximate probability distribution (5.17), we correctly conclude that the population fluctuations are characterized by a mean square magnitude

$$\nu^2 \sim \sigma^2/k_0 = \sigma^2/\Lambda. \qquad (5.18)$$

So long as $\sigma^2$ is appreciably less than $\Lambda$ the fluctuations are not too severe; but once $\sigma^2$ is of the order of, much less greater than, $\Lambda$ the fluctuations are so severe as to disperse the probability cloud to the $n = 0$ axis, so that although the approximation now breaks down it correctly suggests that such a population cannot persist.

## TWO COMPETITORS IN A STOCHASTIC ENVIRONMENT

A next step from single-species populations is to two competing species, with populations $N_1(t)$ and $N_2(t)$.

For specificity, we consider the conventional simple

123

competition equations

$$\frac{dN_1(t)}{dt} = N_1(t) \left[ k_1 - N_1(t) - \alpha N_2(t) \right] \qquad (5.19a)$$

$$\frac{dN_2(t)}{dt} = N_2(t) \left[ k_2 - N_2(t) - \alpha N_1(t) \right]. \qquad (5.19b)$$

Here the coefficient $\alpha$ measures the (symmetric) competition between the two species for the resources measured by the environmental parameters $k_1$ and $k_2$.

In the deterministic case, $k_1$ and $k_2$ are constant, and the analysis proceeds as indicated in Chapter 2. First the deterministic equilibrium populations $N_1^*$ and $N_2^*$ are found by putting the growth rates zero, and then the $2 \times 2$ community matrix in the neighborhood of this equilibrium point is

$$A = \begin{pmatrix} -N_1^* & -\alpha N_2^* \\ -\alpha N_1^* & -N_2^* \end{pmatrix}. \qquad (5.20)$$

Both eigenvalues of this matrix are negative if, and only if, the competition coefficient $\alpha < 1$, which is the well-known Gause-Lotka-Volterra criterion for stable two-species competition. Although derived by a neighborhood stability analysis, the result is easily extended (either analytically, or by topological phase-plane methods) into the entire positive $N_1 - N_2$ population plane. More specifically, we may put $k_1 = k_2 \equiv k_o$, so that

$$N_1^* = N_2^* = k_o/(1 + \alpha) \equiv N^*. \qquad (5.21)$$

The dominant eigenvalue of the community matrix is then $-N^*(1 - \alpha)$, corresponding to the stability determining entity $\Lambda$ being

$$\Lambda = N^*(1 - \alpha). \qquad (5.22)$$

The deterministic stability criterion, $\Lambda > 0$, plainly requires $\alpha < 1$.

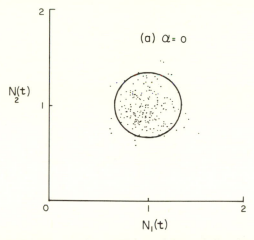

FIGURE 5.5(a). A history of the two populations $N_1(t)$ and $N_2(t)$, obtained by solving the coupled non-linear stochastic differential equations (5.19) with (5.23) on a computer, for a typical run of random numbers. The 200 points represent the population values at successive integer time steps (although the calculation is a continuous time one): they give a good impression of the exact joint probability distribution. The population magnitudes are rescaled so that $k_o = 1$, and we take $\sigma^2 = 0.05$. In this figure the competition coefficient $\alpha = 0$, so that here the two populations are quite independent, the fluctuation character of each being as illustrated in Figures 5.3 and 5.4.

The solid circle derives from the approximate joint probability distribution (5.25), and approximates the probability contour line within which 90% of the points lie. This approximate distribution (5.25) is the two-species analogue of the single-species approximation displayed as the dashed curve in Figure 5.3.

Random environmental fluctuations may be introduced, as before, by considering the parameters $k_1$ and $k_2$ to be random variables. We write

$$k_1 = k_o + \gamma_1(t)$$
$$k_2 = k_o + \gamma_2(t). \tag{5.23}$$

Here $k_o$ is the (common) mean value, and $\gamma_1(t)$ and $\gamma_2(t)$ are independent "white noise" random fluctuations, characterized by a common variance $\sigma^2$, and a common mean

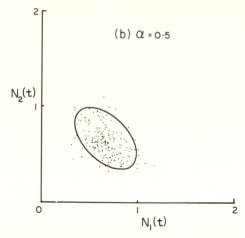

FIGURE 5.5(b). As for Figure 5.5(a), except that the competition coefficient $\alpha = 0.5$, and the two fluctuating populations are now correlated. Again the solid curve denotes the 90% probability contour based on the approximate joint probability distribution (5.25).

of zero. Employing the recipes (5.8) and (5.9), a Fokker-Planck equation may be obtained for the time-independent equilibrium probability distribution, $f^*(n_1, n_2)$. This equation is rather intractable.

However, the exact nonlinear behavior may still be investigated, by studying the original dynamical equations (5.19), with their randomly fluctuating coefficients (5.23), on a computer. In Figure 5.5 we show how the two populations vary together for a typical run of random numbers, for three different values of the competition coefficient $\alpha$, namely $\alpha = 0, 0.5, 0.85$ ($k_o$ and $\sigma^2$ being kept fixed). The nonlinear stochastic differential equations are solved numerically, and the two populations $N_1(t)$ and $N_2(t)$ are represented by a point which is plotted at successive integral time units. (Were we to plot the actual "phase trajectory" as a continuous line, rather than the points at

126

particular times, we would have a mess resembling the tracks of an ink-dipped millipede.) Figure 5.5 is a particular realization of the schematic illustration Figure 5.1(b), and is in effect the two-species analogue of Figure 5.4. The Figures 5.5(a)–(c) clearly show that, as the coefficient $\alpha$ increases toward unity, the relative fluctuations in the populations become more marked.

Although a full analytic solution of the Fokker-Planck equation is not feasible, if the probability cloud is not too diffuse an approximation may be obtained in the spirit of equation (5.17) of the preceding section. Following the previous analysis, we first define quantities $\nu_1$ and $\nu_2$, which measure the relative fluctuations of the two populations about their deterministic mean values, which are given by equation (5.21):

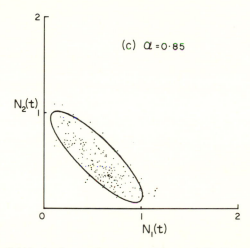

FIGURE 5.5(c). Again as above, but with $\alpha = 0.85$. From the sequence 5.5(a)–(b)–(c) we note that, as the competition coefficient increases, the fluctuations become relatively more and more severe, and the chance of a population fluctuating below any given low level increases. The analytic approximation (5.25) continues to give a reliable impression of the exact joint probability distribution, described by the ensemble of numerically computed points.

$$\nu_1 = \frac{n_1 - N^*}{N^*} \quad \text{and} \quad \nu_2 = \frac{n_2 - N^*}{N^*}. \qquad (5.24)$$

If the stochastic equations (5.19) with (5.23) are now linearized, discarding terms of order $\nu^2$ and higher, before being put into the recipes (5.8) and (5.9) for the Fokker-Planck coefficients, we arrive at a relatively simple equation for the approximate equilibrium probability distribution $\hat{f}^*(\nu_1, \nu_2)$. The solution, obtained in Appendix V, is the bivariate normal distribution

$$\hat{f}^*(\nu_1, \nu_2) = C \exp\left[-\frac{N^*}{\sigma^2}\left(\nu_1^2 + 2\alpha\nu_1\nu_2 + \nu_2^2\right)\right]. \qquad (5.25)$$

Here $C$ is just a normalization constant to make the total probability unity.

The contours of equal probability described by the function (5.25) are ellipses in the plane of the populations $n_1$ and $n_2$. The solid ellipses in Figure 5.5 are the contour lines such that (in this approximation) it is 90% probable the populations $n_1$ and $n_2$ lie inside them. The analytic approximation accords well with the exact numerically derived results in Figure 5.5 (i.e. the exact probability distribution indicated by the dot density); the numerical results suggest the exact probability distribution has contour lines which are somewhat banana-shaped, rather than ellipses.

Alternatively, we may look at the approximate equilibrium probability distribution for any one population in this two-species system. By integrating over all values of the second population, $\nu_2$, we obtain the one-species probability distribution:

$$\hat{f}^*(\nu_1) = D \exp\left[-k_o(1 - \alpha)\nu_1^2/\alpha^2\right]. \qquad (5.26)$$

Here again $D$ is the relevant normalization constant. This result is to be contrasted directly with the result (5.17) for the single-species system. The mean square magnitude of the relative population fluctuation is now roughly characterized by

$$\nu^2 \sim \frac{\sigma^2}{k_0(1 - \alpha)} \sim \frac{\sigma^2}{\Lambda}. \tag{5.27}$$

Recall that the probability cloud will persist for an appreciable time only if this characteristic fluctuation magnitude is significantly less than unity. If we contrast (5.18) and (5.27), the stability provided by the population dynamics is now measured by $k_0(1 - \alpha)$, rather than the single-species value $k_0$; by the eigenvalue magnitude $\Lambda = N^*(1 - \alpha)$ of equation (5.22), rather than the single-species value $\Lambda = N^*$. Thus, as $\alpha$ increases, the interaction dynamics provides an ever-weaker stabilizing influence to offset the randomizing $\sigma$.

These analytic results are approximate, but, as seen in the previous section, they are likely to be qualitatively correct, and they march with the exact numerical work above.

This two-species discussion accords with our previous understanding. In the deterministic case, stability is assured so long as the largest eigenvalue of the interaction matrix remains negative, that is $\Lambda > 0$, which is ensured by $\alpha < 1$. In the stochastic case, even though $\alpha$ remains less than unity, as it increases $\Lambda$ becomes smaller, and the tension between stabilizing interaction dynamics (measured by $\Lambda$) and randomizing environmental fluctuations (measured by $\sigma^2$) is tipped increasingly in favor of the latter.

## PREDATOR-PREY IN A
## STOCHASTIC ENVIRONMENT

The preceding analysis could be repeated, step by step, for a two-species community of predator and prey.

A thorough discussion of such a system has been given

by Morton and Corrsin (1969). They treat a system which is equivalent to a one-predator–one-prey model, with Lotka-Volterra interactions, enriched by a complicated stabilizing density-dependent term in the prey birth rate, and with a term corresponding to environmental stochasticity. On the one hand, the paper gives a penetrating account, from first principles, of the application of the Fokker-Planck equation to the system, and on the other hand, an analogue computer is constructed to check that reality acknowledges this analytic solution.

The detailed results of this model can be shown to re-echo our consistent theme. With no stochasticity, the system is stable, with both eigenvalues of the $2 \times 2$ community matrix having negative real parts ($\Lambda > 0$). With a randomly fluctuating parameter, whose variance is characterized by $\sigma^2$, the prey and predator populations will have fluctuations so severe as soon to produce extinction unless $\Lambda$ is significantly greater than $\sigma^2$.

## MULTISPECIES MODELS IN STOCHASTIC ENVIRONMENTS

For a deterministic environment, we have already seen that the $m$-species population model described by the general set of equations (2.8) will have a stable equilibrium point if, and only if, all $m$ eigenvalues of the community matrix $A$ of equation (2.13) have negative real parts: Figure 5.2(a).

Suppose now that in this general system the environmental fluctuations give rise to random variations in the interaction parameters, such as the birth rates, predation rates, carrying capacities, and so on. The task of finding the equilibrium multivariate probability distribution function $f^*(n_1, n_2, \ldots, n_m)$ for the $m$ fluctuating populations, much less the full time-dependent probability function

$f(n_1, n_2, \ldots, n_m; t)$, is in general clearly quite hopeless.

However, if this $m$-dimensional equilibrium probability cloud is not too diffuse, so that the relative population fluctuations about their *deterministic* mean values,

$$\nu_i = \frac{n_i - N_i^*}{N_i^*} \quad (i = 1, 2, \ldots, m) \qquad (5.28)$$

tend to remain small, the method developed above may be used to get an approximate equilibrium probability distribution. The details of this calculation are outlined in Appendix V. In essence, one observes that, if the white noise in the environmental parameters is relatively small, the equation for the probability distribution function can be expanded around the deterministic equilibrium point. The equilibrium probability distribution, if it exists, is then seen to have the form of a multivariate gaussian distribution in the populations of the $m$ species: the means are the deterministic populations, and the variances and covariances in the populations are related to those in the environmental fluctuations.

Specifically, this approximate equilibrium distribution for the population fluctuations $\nu_1, \nu_2, \ldots, \nu_m$ is the multivariate normal distribution

$$\hat{f}^*(\nu_1, \nu_2, \ldots, \nu_m) = C \exp\left[\sum_{ij} \nu_i B_{ij}^{-1} \nu_j\right]. \qquad (5.29)$$

Here the $m \times m$ symmetric covariance matrix $B$ can be seen to be the solution of the Lyapunov matrix equation

$$D = \tfrac{1}{2}(B\bar{A} + \bar{A}^T B). \qquad (5.30)$$

The elements $d_{ij}$ of the matrix $D$ formally represent the overall covariance between the white noise fluctuations in the stochastic differential equation for the $i$th species and that for the $j$th. The $m \times m$ matrix $\bar{A}$ is

$$\bar{A} = (\mathbf{N}^*)^{-1} A \mathbf{N}^*, \qquad (5.31)$$

131

where $A$ is exactly the community matrix of equation (2.13), evaluated using the average values for all environmental parameters, and $\mathbf{N}^*$ is the $m \times m$ diagonal matrix with diagonal elements $N_i^*$. As observed in Appendix II (p. 193), $\bar{A}$ and $A$ necessarily have identical eigenvalues, so that, as descriptions of the dynamical interactions, $\bar{A}$ and $A$ are effectively the same entity. $\bar{A}^T$ is the transpose of $\bar{A}$. For the equilibrium probability distribution (5.29) to be a sensible one, the quadratic form in the exponent must be negative definite. As $D$ is positive definite, this requires the community matrix $A$ to be a negative form, which is to say to have all its eigenvalues in the left-half plane. Thus the deterministic stability criterion is a necessary, but not sufficient, condition for the existence of an equilibrium probability distribution in the stochastic case.

For the above approximation to be meaningful, we remember that the relative population fluctuations $\nu_i$ must tend to remain significantly less than unity. Expressed as a condition on the matrix $B$, this requires all the eigenvalues of $B^{-1}$ to have magnitudes significantly greater than unity, if the probability cloud is to remain relatively compact. (Remember, $B^{-1}$ is symmetric, so all its eigenvalues are real: it is also necessary, as noted above, that they be negative.) This is just the requirement worked out in detail in one dimension in equation (5.18), and in two dimensions in equation (5.27).

For the rather special case where the matrix $\bar{A}$ is symmetric, as in the class of $m$-species competition models discussed in the next chapter, equation (5.30) simplifies to

$$B^{-1} = D^{-1}\bar{A}. \tag{5.32}$$

If, in addition, the random environmental fluctuations all have a common variance $\sigma^2$, but there are no correlations between the stochastic parameters in the various equations, then $D$ is a diagonal matrix with elements $\sigma^2$. Consequently

$$B^{-1} = (\sigma^{-2})\bar{A}; \qquad (5.33)$$

the eigenvalues of $B^{-1}$ are simply the eigenvalues of $\bar{A}$ (which are identical with those of $A$) divided by $\sigma^2$. The requirement that all the eigenvalues of $B^{-1}$ have magnitudes significantly greater than unity then becomes the requirement that all the eigenvalues of $A$ have magnitudes significantly larger than $\sigma^2$, which is to say the requirement

$$\Lambda > \sigma^2; \qquad (5.34)$$

With this constraint satisfied, the multivariate probability cloud for this particular stochastic model will be compact enough to tend to persist. The corresponding deterministic stability criterion is, of course, merely that

$$\Lambda > 0. \qquad (5.35)$$

## DETERMINISTIC POPULATION VARIATIONS

So far in this chapter, it has been assumed that in the deterministic environment the stable equilibrium populations all have constant (time-independent) values. But a variety of different mechanisms can cause the stable equilibrium community in a fully deterministic environment to exhibit well-defined population oscillations. For such a cyclic equilibrium, the populations of the various species undergo regular, periodic oscillations between maximum and minimum values set by the deterministic environmental parameters in the model equations. Similarly the period of such cycles will be set by the intrinsic parameters of the system.

One way of achieving such a cyclic equilibrium is to allow one or more of the parameters to have some specified periodic time dependence. For example, the model may embody some entirely deterministic seasonality. The simplest paradigm is a single species growing exponentially at a rate $r(t)$ which varies in some periodic fashion:

$$\frac{dN(t)}{dt} = r(t)\, N(t). \qquad (5.36)$$

Here $r(t)$ is some periodic function, with period $T$, having the value zero when averaged over a complete cycle,

$$\int_0^T r(t)\, dt = 0. \qquad (5.37)$$

The consequent population oscillates up and down, with period $T$, in a completely deterministic manner,

$$N(t) = N(o) \exp\left[\int_0^t r(t')\, dt'\right]. \qquad (5.38)$$

Considerably more complicated models can obviously be constructed along these lines, and indeed there exists a large mathematical literature on systems of equations with periodic coefficients.

Stable cyclic equilibria may alternatively be obtained even when all the environmental parameters are unvarying constants. This happens when the nonlinear equations describing the dynamics of interacting populations have stable limit cycle solutions. As we saw in Chapter 4, such deterministic stable limit cycles can easily arise from excessive time delays in feedback mechanisms, or from the dynamics of predator-prey models.

The general tenor of these remarks is that, in a deterministic environment, the community's stable equilibrium is not necessarily a set of constant populations, but may be a well-defined pattern of cyclic population oscillations. Deterministic equilibria may be the point typified by Figure 5.1(a), or equally well the trajectory typified by Figure 5.6(a).

As before, any one of these deterministic environment

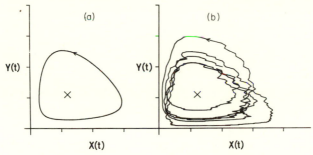

FIGURE 5.6. A specific example of the general points covered in the text. $X(t)$ and $Y(t)$ are dimensionless variables formed from the prey and predator populations, $H(t)$ and $P(t)$, respectively, of equation (4.4): $X = cH$, $Y = (ck/r)P$. In case (a) these population numbers endlessly trace out a deterministic stable limit cycle as time passes; we show computed results for $\beta/b = 1.5$, $cK = 4$, $r = b$. In case (b), with environmental stochasticity introduced via equation (5.39) with $\sigma^2/r_o = 0.1$, we show the populations' trajectory for a typical run of random numbers in (5.39). Figure 5.6(b) gives an impression of the ring-shaped probability cloud that now replaces the deterministic trajectory of 5.6(a).

models may be converted to a stochastic one by allowing one or more of the parameters to have an element of random fluctuation. This continuous spectrum of environmental noise will, as before, destroy the repetitive determinacy of the population cycles. In place of *the* trajectory followed around and around by the populations in Figure 5.6(a), we have a "fuzzed out" probability distribution for the populations, with stochastic fluctuations superimposed on deterministic oscillations.

As a specific example, consider the community comprising a prey population $H(t)$ and predator population $P(t)$, whose dynamics obey the equation (4.4) discussed in Chapter 4 (p. 84). If the prey self-limitation is relatively weak, a stable limit cycle ensues. One such limit cycle for the system is computed in Figure 5.6(a). An element of environmental randomness may be introduced,

135

for example, by writing the prey growth rate parameter $r$ as

$$r = r_o + \gamma(t). \tag{5.39}$$

Here $r_o$ is the (constant) mean value, and $\gamma(t)$ is randomly fluctuating white noise, with mean zero and variance characterized by $\sigma^2$. A typical computer run for the population history in this randomly varying system is shown in Figure 5.6(b), and it conveys a good impression of the donut-shaped probability cloud which now describes the system. The corresponding prey population as a function of time is displayed in Figure 4.3 (p. 91).

In short, there is nothing qualitatively new in the contrast between stochastic and deterministic environments if we begin from a cyclic deterministic equilibrium instead of a constant one. Although the detailed equilibrium behavior can clearly be richly more complicated than the equilibrium point considered throughout the earlier sections, the technical details of the relation between environmentally stochastic and deterministic models retains the character indicated for the stable equilibrium point. Instead of passing from the deterministic point of Figure 5.1(a) to the probabilistic "smoke cloud" of Figure 5.1(b), we pass from the deterministic cycle of Figure 5.6(a) to the probabilistic "smoke ring" indicated by Figure 5.6(b).

## ONE KIND OF SPATIAL HETEROGENEITY

Although the effects of spatial heterogeneity are of the utmost significance in many real populations, such effects are omitted from this book. This was emphasized in the introductory chapter. We pause to lift the ban for a moment.

One way that spatial heterogeneity can influence community dynamics is by admitting a stabilizing balance be-

tween migration and local extinction. In other cases, spatial patchiness or texture can give rise to areas which are
complete or partial refuges for some species, thus contributing to the community stability (Hassell and May,
1973). None of these mechanisms is directly related to the
questions covered in this chapter.

However, on some occasions the spatial heterogeneity
has the effect simply of reducing the effective variance in
the environmental noise. The discussion in this chapter
is then directly relevant. For example, suppose there are
$p$ distinct patches, in each of which the population obeys
the logistic growth equation (5.10), each with its own
stochastic environmental parameter $k$ (whose white noise
spectrum has variance $\sigma^2$), and that there is continuous
dispersal and mingling among the populations of all $p$
patches. The noise level then tends effectively to be $\sigma/p$
in each patch, rather than $\sigma^2$. More generally, we could
consider $p$ patches, each with its own stochastic environment characterized by variance $\sigma^2$, containing communities described by any of the models in this chapter. Then
if the populations keep being completely mixed by migration, again the effective environmental variance is $\sigma^2/p$
rather than $\sigma$. If the mixing is less than complete, the variance per patch is intermediate between $\sigma^2$ and $\sigma^2/p$. These
consequences of this type of spatial heterogeneity have
been pointed out in general terms by Den Boer (1968), and
borne out in detail by the numerical simulations of Reddingius and Den Boer (1970) and Roff (1972).

In such circumstances, if there is no dispersal at all, so
that each patch remains isolated, then in each patch the
interactions between and within species must provide a
stabilization (measured by $\Lambda$) sufficiently great to prevail
against the individual environmental variances $\sigma^2$. If the
environment becomes more variable, the stabilizing dynamical interactions need to become stronger, if the popu-

lation fluctuation level is to remain constant. Thus, with this kind of spatial heterogeneity, migration provides an alternative strategy, whereby the balance between stabilizing population dynamics ($\Lambda$) and destabilizing environmental variance ($\sigma^2$) is preserved not by increasing $\Lambda$ but by effectively decreasing $\sigma^2$.

## SUMMARY

This chapter has dealt with the relation between the dynamics of population models in which all the environmental parameters are strictly deterministic, and the corresponding more realistic models with random environmental fluctuations. The community matrix, evaluated in a deterministic or an average sense as the circumstances dictate, is seen to play a key role in both cases. In the deterministic environment, we need only know the signs of the real parts of the eigenvalues of this matrix to know whether the population interactions are stabilizing. In the stochastic environment, where the system may be in tension between the stabilizing population interactions and the destabilizing environmental fluctuations, we need to contrast the magnitudes of the real parts of these eigenvalues with the magnitudes of the characteristic environmental variances.

# CHAPTER SIX
# Niche Overlap and Limiting Similarity

One of the basic concepts in ecology is the competitive exclusion principle, which forbids the stable coexistence of two or more species making their livings in identical ways. The questions arise, how similar can competing species be if they are to remain in an equilibrium community; how identical is "identical"; how closely can species be packed in a natural environment?

An answer to these questions may begin by noticing that in laboratory experiments, where the environment is carefully kept unvarying, species whose ecology is well-nigh identical have coexisted for long periods (see the reviews by Miller, 1967, pp. 35–46, and 1969, pp. 67–69). A plausible theoretical conjecture is that, in the real world, environmental fluctuations will put a limit to the closeness of species packing compatible with an enduring community, and that species will be packed closer or wider as the environmental variations are smaller or larger (MacArthur, 1971b, 1972; Miller, 1967, p. 67; Slobodkin and Sanders, 1969).

To put the question more sharply, we look at a class of greatly oversimplified models. Consider a one-dimensional resource spectrum, sustaining a series of species, each of which has a preferred position in the spectrum and a characteristic variance about this mean position, described by some "utilization function" as shown in Figure 6.1. For example the resource spectrum may be food size, and the

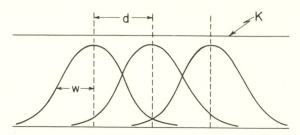

FIGURE 6.1. The curve labeled $K$ represents some resource continuum, say amount of food as a function of food size, which sustains various species whose utilization functions (characterized by a standard deviation $w$ and a separation $d$) are as shown.

consumers may be bird species each having a utilization function which describes their mean food size and its variance. The dynamics of this situation may be plausibly modeled by a system of first-order differential equations, with competition coefficients which depend on how closely species are packed, that is on the degree of niche overlap, on the ratio of $d$ to $w$ in Figure 6.1.

In the stability analysis of such models, two qualitatively different circumstances need be distinguished.

In the unrealistic case when all the environmental parameters are strictly constant (deterministic), then in general the system can remain stable even if an arbitrarily large number of species are packed in, arbitrarily close.

On the other hand, when the relevant environmental parameters fluctuate (stochastic environment), there is an effective limit to the niche overlap consistent with long term stability. This limit to species packing can be shown to depend on the environmental variance in a far-from-obvious and extremely interesting way (cf. Figure 6.3). If the fluctuations in the resource spectrum are severe, having variances comparable in magnitude with their mean values, the species packing is indeed roughly proportional to the environmental variance, as one would expect intuitively. But, for fluctuations ranging from moderate to exceedingly small, the species packing attains an effective limiting value roughly equal to the width of the utilization functions. Thus as the ratio between the variance and mean value in the resource spectrum or other pertinent environmental parameter falls from 0.3 to 0.0001, the closest species packing consistent with stability falls only from 2 to 1 times the utilization function variance. Moreover, our general result is a robust one, being rather insensitive to the details of the mathematical model.

Collecting these statements, we observe that the species packing parameter $d$ indeed goes to zero when the environmental variance becomes strictly zero, but that for any finite environmental variance, $d$ remains roughly equal to the utilization function width, $w$. This result, which at first glance seems odd, reflects the technical fact that, when a large number of species compete on a one-dimensional resource continuum, the mathematics describing the stability contains an essential singularity around $d = 0$ (equations (6.18) and (6.22), and Figure 6.2). Thus there is here a qualitative difference between an environmental variance which is small but finite and one which is zero. Although not possessing exactly the essential singularity of the many-species limit, the results for 3- or 4-species competition are for most practical purposes indistinguishable from it (Figures 6.2 and 6.3).

Following Hutchinson's (1959) initial observations, MacArthur (1971b, 1972) has recently reviewed a body of semiquantitative field work bearing on species packing and character displacement among competing species. This empirical data, which is discussed more fully below, marches with the theoretical suggestion that there can be an effective limit to niche overlap in the real world, and that this limit is relatively insensitive to the degree of environmental variability (unless it is very large).

We now go on to define and discuss a rather special model for competition on a one-dimensional resource continuum. Various directions in which these results may be generalized are then indicated. Particular attention is paid to enlarging the number of relevant niche dimensions, and to the relation between the shape of the resource spectrum and the number of species present in the equilibrium community. It is finally emphasized, inter alia, that the discussion pertains to situations dominated by competition within one trophic level, which tends to restrict the circumstances in which theory confronts reality.

## A "MICROSCOPIC" MODEL FOR
## ONE-DIMENSIONAL COMPETITION

Consider a one-dimensional continuum of resources, such as food size, or vertical habitat, or horizontal habitat. This may be schematically depicted as in Figure 6.1, where the curve labeled $K(x)$ shows amount of food as a function of food size, or amount of habitat as a function of height, and in general amount of resource $K$ as a function of some relevant parameter $x$. Suppose further that this resource sustains various species, each of which has a utilization function $f_i(x)$ as depicted in Figure 6.1, which characterizes the species' use of the resource spectrum: the subscript $i$ labels the species. In particular, we note the mean position and the standard deviation, $w$, about this mean for the various species; i.e. the mean and the variance of the food size, or of the habitat height, etc. The separation, $d$, between the mean positions of species which are adjacent on the resource continuum will clearly be a measure of how densely the species are packed.

In an interesting series of papers, MacArthur (1969, 1970, 1972) has established a criterion which ensures that the actual community utilization of the resource will pro-

142

vide the best least-squares fit to the available resource spectrum. This requires the populations of the $m$ species, $N_i(t)$ [labeled sequentially $i = 1, 2, \ldots, m$], to obey

$$\frac{dN_i(t)}{dt} = N_i(t) \left[ k_i - \sum_{j=1}^{m} \alpha_{ij} N_j(t) \right], \qquad (6.1)$$

where the $k_i$ are integrals with respect to $x$ over the product of the resource spectrum and the utilization function of the $i$th species, and the competition coefficients $\alpha_{ij}$ are convolution integrals between the utilization functions of the $i$th and $j$th species. With this, we are assured both that the equilibrium populations (obtained by setting all $d/dt = 0$) minimize the squared difference between available and actual "production," and also that nonequilibrium initial populations will move in time toward this minimum configuration.

Equation (6.1) is of course the Lotka-Volterra competition equation, but tied now to the underlying "microscopic" model illustrated by Figure 6.1, so that we have explicit recipes for the $k_i$ and $\alpha_{ij}$ in terms of direct biological assumptions. Previously the "macroscopic" parameters $k_i$ and $\alpha_{ij}$ were phenomenological constants.

Specifically, we may represent the squared difference between actual and available production by $Q$,

$$Q(t) = \int \left[ K(x) - \sum_i f_i(x) N_i(t) \right]^2 dx. \qquad (6.2)$$

Here the total $Q$ is the squared difference at each point along the one-dimensional resource continuum, integrated over the continuum. It may readily be seen that the constant (equilibrium) values of the populations which minimize this expression, $N_i^*$, obey the set of algebraic equations

143

$$k_i - \sum_j \alpha_{ij} N_j^* = 0, \qquad (6.3)$$

where the quantities $k_i$ and $\alpha_{ij}$ are defined explicitly by

$$k_i = \int K(x) f_i(x)\, dx, \qquad (6.4)$$

$$\alpha_{ij} = \int f_i(x) f_j(x)\, dx. \qquad (6.5)$$

With this definition, $Q$ may be neatly rewritten as

$$Q(t) = Q_o + \sum_{i,j} (N_i(t) - N_i^*) \alpha_{ij} (N_j(t) - N_j^*), \qquad (6.6)$$

where $Q_o$ is the (constant) minimum value.

We may thus visualize that, just as a loudspeaker (approximately) synthesizes complicated sounds by superposing pure sinusoidal Fourier components each with its appropriate amplitude, so here the resource spectrum $K(x)$ is (approximately) synthesized from the addition of the $m$ "generalized Fourier components" $f_i(x)$, with the populations $N_i$ being the "Fourier coefficients." The choice of the equation (6.1) ensures both that the equilibrium populations give the best least-squares fit (i.e. are the optimal "generalized Fourier amplitudes") and that any other choice of populations will tend to this equilibrium as optimal. That is, $dQ/dt \le 0$, with the equality pertaining only if all $N_i = N_i^*$, a fact which is easily verified by differentiating (6.6) with respect to $t$ and using (6.1).

A more detailed account of the relation between the quantities $k_i$ and $\alpha_{ij}$ in the dynamical equation (6.1), and their expression in terms of a model of resource utilization, takes into consideration that utilization is not just percentage of time or percentage of diet, but rather has weighting terms for resource renewal ability. For such a fully detailed

144

presentation, see the work of MacArthur (1968, 1970, 1971b, 1972, Chs. 2 and 7). However, the above account is correct in essentials, and gives the right recipe for the elements in the coefficient matrix, upon which the system stability devolves.

Given any particular resource spectrum, $K(x)$, and set of utilization functions, $f_i(x)$, we first calculate the quantities $k_i$ and $\alpha_{ij}$ from (6.4) and (6.5), and then substitute into equation (6.3) to identify the equilibrium populations. To be biologically meaningful, all the $N_i^*$ must be positive; any populations with negative $N_i^*$ will be absent from the equilibrium community. This relation between the shape of the resource spectrum and the species composition at equilibrium already provides one rough line of attack upon the species packing problem, and we shall return to it below.

Having found the equilibrium point, the next step is to see whether it is stable to small perturbations to the populations. Such a stability analysis, as usual, hinges on the eigenvalues of the community matrix $A$, whose elements here are clearly (cf. equation (2.13))

$$a_{ij} = -N_i^* \alpha_{ij}. \tag{6.7}$$

The system will tend to return to equilibrium, with small perturbations tending to die away, if and only if all the eigenvalues of this matrix lie in the left half of the complex plane.

For specificity, we begin by assuming that in the equilibrium community all populations are equal, $N_i^* = N^*$; for a large number of species present, $m \gg 1$, this means a flat resource spectrum. Then $a_{ij} = -N^* \alpha_{ij}$, and the stability of the equilibrium depends simply on the eigenvalues of the competition matrix $\boldsymbol{\alpha}$, whose elements are $\alpha_{ij}$: all these eigenvalues need to lie in the right half complex plane.

If this $\boldsymbol{\alpha}$ matrix is symmetric, as it often will be, all its eigenvalues are necessarily real. Writing $\lambda_{\min}$ for the smallest eigenvalue of the $\boldsymbol{\alpha}$ matrix, this deterministic stability criterion is

$$\lambda_{\min} > 0. \tag{6.8}$$

Throughout the above formal presentation, it has been taken for granted that the environment is unvarying, deterministic. In reality there will, to a greater or lesser degree, be random fluctuations in the resource spectrum, and thus in the environmental parameters $k_i$. In the spirit of Chapter 5, we assume the random noise in the resource continuum can be represented by writing

$$k_i = \bar{k}_i + \gamma_i(t). \tag{6.9}$$

Here $\bar{k}_i$ is a constant, being the mean value (having the common value $\bar{k}$ for large $m$), and the random variables $\gamma_i(t)$ are white noise, with mean zero and variance measured by $\sigma^2$. It is further assumed that there is no covariance between $\gamma_i(t)$ and $\gamma_j(t)$. These assumptions are reviewed below.

Following the discussion toward the end of the preceding chapter, we note that in such a randomly fluctuating environment the probability of a species becoming extinct will be small (corresponding to the mechanical "stability" of the deterministic case) if the smallest eigenvalue of the competition matrix $\boldsymbol{\alpha}$ very roughly obeys

$$\lambda_{\min} > \sigma^2/\bar{k}. \tag{6.10}$$

This result is commonsensical. In a randomly fluctuating environment, it is not enough that all the eigenvalues be positive, but rather they should be bounded away from zero by an amount roughly proportional to the environmental noise level.

We now proceed to see how these stability ideas relate to the permissible degree of niche overlap.

## THE COMPETITION MATRIX

A reasonable assumption for the detailed shape of the species' utilization function is the familiar bell-shaped gaussian,

$$f_i(x) = C \exp\left[-x^2/(2w_i^2)\right]. \tag{6.11}$$

Here $C$ is a normalization constant, and $w_i$ is the width of the utilization function, as illustrated in Figure 6.1. From equation (6.5), the competition coefficients are immediately

$$\alpha_{ij} = (w_i w_j \pi)^{-1/2} \int_{-\infty}^{\infty} \exp\left[-\frac{x^2}{2w_i^2} - \frac{(x - d_{ij})^2}{2w_j^2}\right] dx$$

$$= C_{ij} \exp\left[-d_{ij}^2/2(w_i^2 + w_j^2)\right]. \tag{6.12}$$

Here $d_{ij}$ is defined to be the distance along the resource spectrum from the mean position of the $i$th species to that of the $j$th, and $C_{ij}$ is the normalization constant $(2w_i w_j/(w_i^2 + w_j^2))^{1/2}$; notice that if $w_i = w_j$, $C_{ij} = 1$.

In particular, if the species' utilization functions have a common width $w$, and are uniformly spaced along the resource continuum a distance $d$ apart, so that $d_{ij} = (i - j)d$, as in Figure 6.1, the competition coefficients are

$$\alpha_{ij} = \alpha^{(i-j)^2} \tag{6.13}$$

with, for notational convenience, the definition

$$\alpha \equiv \exp\left[-d^2/4w^2\right]. \tag{6.14}$$

The quantity $\alpha$ is small if there is little overlap, $d/w \gg 1$, but tends to unity for substantial overlap, $d/w < 1$. The competition matrix $\boldsymbol{\alpha}$ for an $m$-species community with such competition coefficients is

$$\boldsymbol{\alpha} = \begin{pmatrix} 1 & \alpha & \alpha^4 & \alpha^9 & . & \alpha^{m^2} \\ \alpha & 1 & \alpha & \alpha^4 & . & . \\ \alpha^4 & \alpha & 1 & \alpha & . & . \\ \alpha^9 & \alpha^4 & \alpha & 1 & . & . \\ . & . & . & . & . & . \\ \alpha^{m^2} & . & . & . & . & 1 \end{pmatrix} \cdot \qquad (6.15)$$

*Deterministic Environment*

The matrix (6.15) is a positive-definite form for all $0 \leqslant \alpha < 1$, that is for all $d$ (see equation (6.14)). All its eigenvalues are necessarily positive. Indeed this is a general property of *any* competition matrix whose elements are generated by (6.5), as is proved by observing that the general quadratic form

$$\Phi = \sum_{i,j} y_i \alpha_{ij} y_j$$

$$= \int \left[ \sum_i f_i(x) y_i \right]^2 dx \qquad (6.16)$$

is necessarily positive.

Thus the criterion (6.8) is always satisfied, with the consequence that stability sets no limit to the species packing in a strictly deterministic environment. This is the result noted at the beginning of the chapter. Moreover in general the more species packed in, the better the least-squares fit to the deterministic resource spectrum.

Even so, it is interesting to see how the *smallest* eigenvalue of $\boldsymbol{\alpha}$, which sets the stability character, varies with niche overlap, as measured by $d/w$. For $m \gg 1$, we have (see Appendix II)

$$\lambda_{\min} = 1 - 2\alpha + 2\alpha^4 - 2\alpha^9 + 2\alpha^{16} \ldots \qquad (6.17)$$

148

This series may be summed (May and MacArthur, 1972), to get an approximation which is very accurate unless $d \gg w$:

$$\lambda_{\min} = 4\pi^{1/2}(w/d) \exp\left[-\pi^2 w^2/d^2\right]. \tag{6.18}$$

(The corrections to this approximation do not exceed 1% until $d/w > 4.1$.)

This is a remarkable result. For substantial niche overlap, i.e. $d/w$ small, $\lambda_{\min}$ tends to zero faster than any finite power of $d/w$: there is an essential singularity at $d/w = 0$. Thus although $\lambda_{\min}$ is indeed necessarily positive even for small $d/w$, it becomes exceedingly tiny, corresponding to extremely long damping times. This foreshadows the stochastic environment results.

In the case of a relatively small number of competing species, namely $m = 2, 3, 4$, the smallest eigenvalue of the competition matrix may be found by straightforward algebraic manipulation of the relevant determinantal equation (May, 1972g).

In general it may be seen that, for very little niche overlap, $d \gg w$, $\alpha \to 0$, and all the minimum eigenvalues tend to unity. Conversely, for appreciable niche overlap, $d < w$, we have $\alpha \to 1$ and the minimum eigenvalues become increasingly tiny as the number of species increases. In the limit $d \ll w$ we have (May, 1972g)

$$\lambda_{\min}(m = 2) \to (d/2w)^2 \tag{6.19}$$

$$\lambda_{\min}(m = 3) \to \frac{4}{3}\left(\frac{d}{2w}\right)^4 \tag{6.20}$$

$$\lambda_{\min}(m = 4) \to \frac{12}{5}\left(\frac{d}{2w}\right)^6 \tag{6.21}$$

while for $m \gg 1$ the dependence is as given by (6.18).

Figure 6.2 illustrates these minimum eigenvalues as functions of niche overlap, $d/w$, for communities of 2, 3, 4,

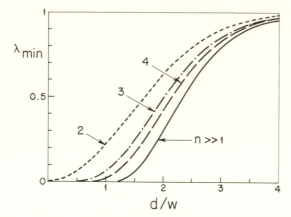

FIGURE 6.2. The minimum eigenvalue of the competition matrix (6.15) (based on gaussian utilization functions) as a function of niche overlap, $d/w$, for a sequence of $n$ species competing on a one-dimensional resource continuum. Here $n = 2, 3, 4$ and $n \gg 1$.

and $m \gg 1$ competing species. Notice that for practical purposes, $m = 4$ is hard to distinguish from "$m = \infty$."

### Stochastic Environment

If the resource spectrum is subject to random fluctuations, the overall community is in tension between the stabilizing dynamical interactions between and within species (measured by the smallest eigenvalue of the competition matrix, $\lambda_{\min}$), and the destabilizing random environmental fluctuations (measured by the characteristic relative variance, $\sigma^2/\bar{k}$). Long-term existence of the community then requires that the stability criterion (6.10) be roughly fulfilled.

Combining the qualitative equation (6.10) with the results for $\lambda_{\min}(m)$ summarized by Figure 6.2, we arrive at an estimate of the closest species packing, $d/w$, consistent with stability for a given environmental noise level, $\sigma^2/\bar{k}$. These results, illustrated in Figure 6.3, are as discussed at the start of the chapter.

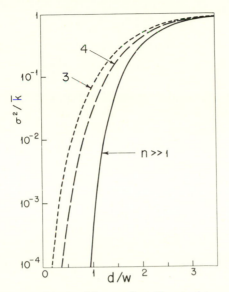

FIGURE 6.3. The closest niche overlap, $d/w$, consistent with community stability in a randomly varying environment, whose fluctuations are characterized by a variance (relative to the mean) of $\sigma^2/\bar{k}$. The variance is plotted on a logarithmic scale to emphasize that, over a wide range, it has little influence on the species packing distance for $n > 2$. This figure is based on the gaussian utilization functions of Figure 6.2.

In particular we see explicitly from equation (6.18) that for large $m$ this closest degree of niche overlap is $d \sim w$, depending on the environmental fluctuations only as $\sqrt{\ln \sigma^2}$, which is a very weak dependence. The results for $m = 3, 4$, although allowing a slightly closer limiting packing distance, display a similar insensitivity to the degree of random fluctuation, so long as it is not severe.

## GENERALIZATIONS OF THE MODEL

The question arises, to what extent are the above results peculiar to the particular model? We catalogue some answers.

*Other Utilization Functions*

The previous analysis may be repeated for various alternative utilization functions $f(x)$, ranging from more "fat-tailed" to more sharply cut off than the gaussian (6.11). This leads to competition matrices different from (6.15). However, for $m \gg 1$ the smallest eigenvalue $\lambda_{min}$ in general retains its essentially singular behavior as $d/w$ becomes small (May, 1972g), thus preserving the basic character of the earlier results.

Indeed it can be shown that *any* well-behaved continuous utilization function $f(x)$ leads in the many-species limit, $m \gg 1$, to a competition matrix whose minimum eigenvalue has an essential singularity as $d/w \to 0$:

$$\lim_{d/w \to 0}(\lambda_{min}) \to (2cw/d) \exp\left[-2\pi\xi_o w/d\right]. \quad (6.22)$$

Here $\xi_o$ and $c$ are pure numbers, of order unity. (In detail, if the analytic function $f(z)$ in the complex plane has its pole with smallest imaginary magnitude at $z_o = w(\zeta_o + i\xi_o)$, thus defining the number $\xi_o$, then $c$ is the remarkable number $|2\pi \text{ Residue } f(z_o)|^2$. This and associated other results are set out fully in May (1972g); the mathematics is unlikely to be of wider interest outside this specific context, and we do not develop it here.)

The elegant general result (6.22), in conjunction with the crude criterion (6.10) for stability in a fluctuating environment, implies an effective limit to niche overlap and species packing at around $d \sim w$, with only a weak (logarithmic) dependence on the environmental noise level $\sigma^2$. Our earlier results are thus not pathological artifacts of the gaussian utilization functions.

*Nonlinear Stability Analysis*

The above stability analysis was presented, particularly in the deterministic case, as a linearized neighborhood analysis. But, as pointed out in Chapter 3 (p. 55), the quantity $Q(t)$ of equation (6.2) or (6.6) is in fact a Lyapunov function for the system: $Q \geq 0$ and $dQ/dt \leq 0$ throughout the domain of positive populations, with the equalities pertaining to the equilibrium configurations. Consequently the full nonlinear stability character is legitimately described by the neighborhood analysis.

### Spacing and Widths of Utilization Functions

The earlier work assumed that the utilization functions share a common shape, with common width, and that they are uniformly spaced along the resource continuum measured by $x$, the food size or habitat height or whatever. However, as is made particularly clear in the derivation of the general theorem (6.22), all we really need is that the niche overlap *ratio*, $d/w$, be uniform.

Indeed, as emphasized by MacArthur (1971b, 1972, p. 65), if the resource spectrum in question is food size, as measured by food weight or length, there are general reasons why one would expect a uniform sequence on a logarithmic scale of food size. This is borne out by the work reviewed by MacArthur (Storer, 1966; Hespenheide, 1971) and that of Schoener and Gorman (1968). In other words, we expect and find that species competing on a one-dimensional food resource continuum tend to sort out in a sequence such that successive weights are a constant multiple (as first observed by Hutchinson, 1959), rather than a constant difference: a geometric sequence rather than an arithmetic one.

In short, if the width $w$ changes in some systematic way along the resource continuum, our results are preserved so long as the separation $d$ changes in the same proportion, keeping $d/w$ roughly constant. Thus the theory is directly applicable to the circumstances of the preceding paragraph.

In passing, we remark that it would be nice to explore models in which the niche spacings $d$ and widths $w$ themselves included an element of stochasticity. This would introduce some of the features met in the "randomly connected food webs" of Chapter 3, with (to a degree) random spacing of utilization functions of random widths. I conjecture that the essentials of the above models would remain, but this is an interesting open question.

### The Model for the Population Dynamics

As discussed earlier, the parameters $k_i$ and $\alpha_{ij}$ in the "macroscopic" population equations (6.1) can be linked to a "microscopic" model involving the shape of the resource spectrum and of the various species' utilization functions, by MacArthur's (1969, 1970, 1972) idea that the equilibrium community is the one which optimizes the least-squares fit between actual production and that available.

Quite apart from the teleology implicit in assuming that communities minimize anything, a choice of fit other than least-squares will lead to equations superficially different from (6.1). However, their competition matrix characterizing small displacements from equilibrium will end up similar to those above. As mentioned in Chapter 3 (p. 54), and emphasized by many people, equation (6.1) represents the first term in a Taylor expansion of a much wider class of equations, and thus should be useful in discussing the stability in the neighborhood of equilibrium configurations.

It may be added that, although symmetry of the competition matrix ($\alpha_{ij} = \alpha_{ji}$) is necessary for some of MacArthur's results, and although our assumption of a common shape for all utilization functions implies such symmetry via equation (6.5), the main result, namely the essential singularity in the smallest eigenvalue of the competition matrix, does not require such symmetry. For example, suppose the convolution integral (6.5) for the $\alpha_{ij}$ were different going up the spectrum $(i < j)$ from going down it $(i > j)$, as Roughgarden (1972) argues it often is in reality (see his Figure 3). The ensuing competition matrices are usually amenable to the same methods which led to the general result (6.22), and imply a similar essential singularity (May, 1972g).

## The Environmental Stochasticity

The qualitative criterion (5.34), whence (6.10), for a community in a randomly fluctuating environment to persist, rather than to suffer extinction, assumed the stochasticity in the $k_i$ parameters to be white noise, that is not correlated between successive time intervals. It also assumed the fluctuations in $k_i$ to be uncorrelated with those in $k_j$.

Insofar as there are correlations, either in time for any one $k_i$ or along the resource continuum between $k_i$ and $k_j$, they act such as effectively to diminish the environmental variance $\sigma^2$.

Were our broad general conclusions sensitively dependent on the value of $\sigma^2$, this would be worrying, because such correlations clearly are present to a certain extent in real situations. But as our results are in fact quite insensitive to $\sigma^2$, unless it be severe, the idealization of assuming no correlations in the noise spectra should give an adequate characterization of reality.

## SHAPE OF THE RESOURCE SPECTRUM

Previous theoretical studies of "limiting similarity," or limiting degree of niche overlap, have focused on the relation between the shape of the resource spectrum and the number of non-zero populations with given utilization functions which can be fitted onto it (MacArthur and

Levins, 1967; Levins, 1968a; Rescigno, 1968; Vandermeer, 1970, 1972; Roughgarden, 1973 and private communication; for a lucid summary see MacArthur, 1972, Appendices to Chs. 2 and 8). This work constitutes a *deterministic* analysis of the possible equilibrium configurations. As noted above, such a deterministic equilibrium is necessarily stable within the present general framework. If, however, this deterministic analysis requires the shape of the resource spectrum to lie within very narrow limits in order for all competing species to be present in the equilibrium community, it may be plausibly argued that environmental vagaries in the real world will upset such an equilibrium.

Such verbal arguments, which have been cogently presented by Miller (1967) and others as well as the above mentioned authors, clearly have validity. Their shortcoming is that they are incapable of saying anything precise about the relation between limiting niche overlap and degree of environmental variability.

The relation between the resource spectrum shape and the equilibrium community is now pursued in further detail, partly for its own sake, and partly because it is relevant to the relaxation of our earlier convenient assumption that all equilibrium populations were equal.

The $m$ equilibrium populations for the dynamical equations (6.1) are given by the set of $m$ linear equations (6.3), which may be rewritten

$$\mathbf{N}^* = \boldsymbol{\alpha}^{-1}\mathbf{k}. \tag{6.23}$$

Here $\mathbf{N}^*$ is the $m \times 1$ column matrix of the equilibrium populations, and $\mathbf{k}$ the column matrix of the parameters $k_i$ which summarize the shape of the resource spectrum. For specificity, we use the biologically sensible gaussian utilization functions, so that the competition matrix $\boldsymbol{\alpha}$ is the $m \times m$ matrix (6.15). As MacArthur (1972, p. 42) has remarked, this form for the competition coefficients should be a good approximation even if the generating utilization functions are not gaussian.

Consider first a two-species system. Then (6.23) gives

$$N_1^* = (k_1 - \alpha k_2)/(1 - \alpha^2) \qquad (6.24\text{a})$$

$$N_2^* = (k_2 - \alpha k_1)/(1 - \alpha^2). \qquad (6.24\text{b})$$

For the coexistence of the two species in equilibrium, we require $N_1^* > 0$ and $N_2^* > 0$, leading to the constraint on the resource spectrum shape (remember that necessarily $\alpha < 1$):

$$\frac{1}{\alpha} > \frac{k_1}{k_2} > \alpha. \qquad (6.25)$$

Equation (6.14) expresses $\alpha$ in terms of the niche overlap parameter, $d/w$. Figure 6.4 illustrates the range of resource spectrum shapes, characterized by the ratio $k_1/k_2$, which admit of a two-species equilibrium, as a function of the degree of niche overlap. The dashed line shows the specific shape which leads to the two equilibrium populations being equal, namely in this case obviously $k_1 = k_2$.

For a three-species system, with shape parameters $k_1$, $k_2$, and $k_3$, it would be nice to display as functions of $d/w$ the two-dimensional surfaces of $k_1/k_2$ and $k_1/k_3$ values which bound the domains wherein all three species coexist in the deterministic equilibrium community. Since this page is two-dimensional, it is easier to show a slice through the domain, and this we do by putting $k_1 = k_3$ (corresponding to a symmetrical resource spectrum). Then

$$N_1^* = N_3^* = (k_1 - \alpha k_2)/(1 - \alpha^2)^2 \qquad (6.26\text{a})$$

$$N_2^* = (k_2[1 + \alpha^4] - 2\alpha k_1)/(1 - \alpha^2)^2, \qquad (6.26\text{b})$$

leading to the criterion for a three-species equilibrium

$$\frac{1 + \alpha^4}{2\alpha} > \frac{k_1}{k_2} > \alpha. \qquad (6.27)$$

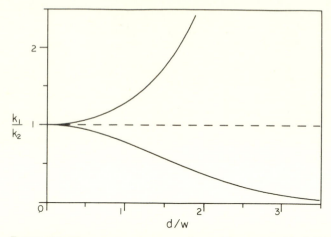

FIGURE 6.4. The resource spectrum shapes (characterized by $k_1/k_2$) which allow a 2-species equilibrium configuration, as a function of degree of niche overlap, $d/w$. The permissible range of shape parameters lies between the two solid curves. The dashed line indicates the shape which equalizes the two equilibrium populations. See text for further details.

This is illustrated in Figure 6.5, which for this three-species example is the analogue of Figure 6.4 for the two-species system. Again the dashed line depicts the particular shape which gives all the equilibrium populations equal, as assumed throughout the earlier sections.

Similar explicit pictures can be drawn for other slices through the $k_1 : k_2 : k_3$ parameter space for a three-species assembly; through the $k_1 : k_2 : k_3 : k_4$ parameter space for four species; and through the $k_1 : k_2 : k_3 : k_4 : k_5$ region for five species. The general features and trend evidenced by Figures 6.4 and 6.5 continue to be manifested by such figures. By way of one among many possible further illustrations, we show the range of shape parameters which permit an equilibrium five-species community, as a function of degree of niche overlap. Here the particular slice through the parameter hypervolume is the symmetrically

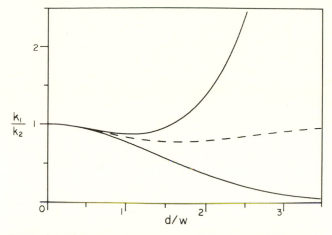

FIGURE 6.5. The range of resource spectrum shapes which permits a 3-species equilibrium community, as a function of niche overlap, $d/w$. Here we assume a symmetric spectrum, $k_1 = k_3$, so that a one-parameter shape characterization $(k_1/k_2)$ is possible. Again the dashed line indicates the condition where all 3 equilibrium populations are equal.

undulatory shape with $k_1 = k_3 = k_5$ and $k_2 = k_4$, so that a one-dimensional characterization in terms of $k_1/k_2$ is feasible. The relevant criterion illustrated by Figure 6.6 is

$$\frac{(1 + \alpha^4)^2}{\alpha(2 + \alpha^4 + \alpha^8)} > \frac{k_1}{k_2} > \frac{2\alpha(1 + \alpha^2)(1 + \alpha^4 + \alpha^8)}{(1 + \alpha^4)(1 + 2\alpha^2 + \alpha^4 + \alpha^6 + \alpha^{10})}.$$

$$(6.28)$$

The hatched region in Figure 6.6 is that where an equilibrium five-species community exists (there being none for large overlap, $d/w < 0.78$, for this particular shape), and the slightly larger region, extending down to $d/w \to 0$, is that where an equilibrium four-species assembly, purged of the central population $N_3$, can exist.

The overall features of Figures 6.4, 6.5, and 6.6 are clear. So long as neighboring utilization function's means are separated by a distance greater than their intrinsic width,

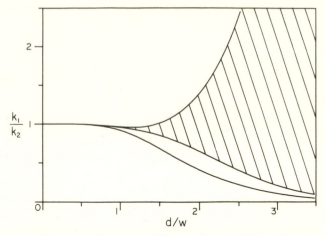

FIGURE 6.6. One of many possible extensions of Figures 6.4 and 6.5 to communities with more than 3 species. The cross-hatched region (which begins at $d/w = 0.78$) indicates the range of symmetrically wavy resource spectrum shapes ($k_1 = k_3 = k_5$; $k_2 = k_4$) which permit a 5-species equilibrium community. We display the range of shape parameters, $k_1/k_2$, as a function of niche overlap, $d/w$. The wider range of $k_1/k_2$, bounded by the upper and lower solid lines and extending to $d/w = 0$, allows an equilibrium 4-species community (with $N_3$ missing) for this particular shape. See text for details.

multispecies equilibria are possible for a wide range of resource continuum shapes. This region of wide tolerance coincides with the region where it was previously shown in detail that equilibria are stable with respect to random fluctuations in the resource spectrum, unless such fluctuations are severe (Figure 6.3). Conversely, in the region of large niche overlap ($d$ smaller than $w$), where it has previously been seen that equilibrium communities are unstable to even a very small amount of environmental fluctuation, we notice that a narrowly specific range of resource spectrum shapes is required to allow the initial equilibrium to be possible. As Figures 6.4, 6.5, and 6.6 bear out, this trend is manifested with increasing sharpness as the num-

160

ber of competing species in the community increases. The general features are, however, already well established for $m = 3$.

It is of course natural that the earlier discussion of stochastic stability and the present section on existence of deterministic equilibria are so connected, as both ultimately devolve on the structure of the competition matrix.

Although our earlier stability analysis assumed that in equilibrium all populations were equal, the considerations of this section suggest that our results are not strongly dependent on this feature, so long as all species are present in significant numbers in the equilibrium community.

The qualitative conclusion emerging from these results, both for the equilibrium configuration and for its stability in a randomly fluctuating environment, is that communities in which many species compete on a one-dimensional resource continuum are robust for $d > w$, fragile for $d < w$.

## COMPETITION IN MORE-THAN-ONE DIMENSION

The preceding discussion has been wholly for competition in one resource dimension.

Cody (1968) has classified the partitioning in the three resource dimensions of horizontal habitat, vertical habitat, and food for ten grassland bird communities around the world. Although his purpose was mainly to study some of Hutchinson's (1957) ideas about niche hypervolume, we may pervert Cody's results to our present purpose by observing they show eight of the communities to be organized largely in one dimension (food selection). Thus there are circumstances where our one-dimensional model is at least a good first approximation.

Once there are two or more niche dimensions, the utilization functions depend on two or more independent variables, $x$, $y$, . . . , which label the various resource continua: $f(x, y \ldots)$. The consequent competition coefficients will be calculated, as multidimensional integrals, from the appropriate generalization of (6.5).

Two limiting cases may be distinguished. These "bracket" the general case.

### Orthogonal Dimensions

One limiting case arises when the pertinent resource dimensions are completely independent, which alternatively may be expressed by saying the niche dimensions are orthogonal. This means the utilization functions may be factored into their separate resource components,

$$f(x, y \ldots) = f_1(x) f_2(y) \ldots \tag{6.29}$$

The consequent integrals for the competition coefficients are separable, and the competition coefficients also factor:

$$\alpha_{ij} = (\alpha_1)_{ij}(\alpha_2)_{ij} \ldots \tag{6.30}$$

Here $(\alpha_k)_{ij}$ describes the competition between species $i$ and $j$ in the $k$th resource dimension.

In this rather extreme circumstance, the various dimensions may be treated separately, one by one, with the previous analysis being applicable to each dimension individually. The point is developed somewhat more fully by May (1972g). The earlier results thus remain intact in their essentials.

### Simplex

The opposite extreme limit arises when every species interacts with every other species in a manner which is completely symmetrical among the $n$ niche dimensions.

To envisage this more explicitly, begin with two species competing on a one-dimensional resource continuum. The competition matrix is

$$\boldsymbol{\alpha} = \begin{pmatrix} 1 & \alpha \\ \alpha & 1 \end{pmatrix}, \tag{6.31}$$

where $\alpha$ measures the interaction as before (equation (6.5)). Now add a second resource dimension, and a third species. Symmetric competition in this two-dimensional resource space may be conceptualized by thinking of the three utilization functions as three overlapping circles, with their centers at the points of some equilateral triangle: each circle overlaps the other two in a symmetrical fashion, and the degree of overlap measures the magnitude of the two-dimensional competition coefficient $\alpha$. The competition matrix is now

$$\boldsymbol{\alpha} = \begin{pmatrix} 1 & \alpha & \alpha \\ \alpha & 1 & \alpha \\ \alpha & \alpha & 1 \end{pmatrix} \tag{6.32}$$

For three resource dimensions and four species, we imagine, as it were, four fuzzy interpenetrating cannonballs, whose centers form a regular tetrahedron. Again the symmetrical overlap measures the competition coefficient $\alpha$, and the $4 \times 4$ competition matrix again has "1" down the diagonal, and "$\alpha$" everywhere else.

Images fail us when we move on to more than three resource dimensions, but the mathematics is straightforward as we ascend the hierarchy of $n$-dimensional "simplexes" which follow on from the equilateral triangle and the regular tetrahedron. For $n + 1$ species competing in this fashion in an $n$-dimensional resource space, the $(n + 1) \times (n + 1)$ competition matrix $\boldsymbol{\alpha}$ is the obvious extension of equations (6.31) and (6.32), with again "1" down the diagonal and "$\alpha$" elsewhere. This matrix has an $n$-fold degenerate smallest eigenvalue, namely

$$\lambda_{\min} = 1 - \alpha. \tag{6.33}$$

In the limit of substantial niche overlap, this minimum eigenvalue will scale approximately as $d^2/w^2$ (following equation (6.14)), where we identify $d$ and $w$ as providing some measures of the niche separation and intrinsic niche width, respectively, in this symmetrical $n$-dimensional circumstance: $d$ is the distance between cannon-ball centers, $w$ the cannon-ball radius. The rough stability criterion (6.10) then leads to the qualitative requirement that

$$d > w(\sigma^2/\bar{k})^{1/2}. \tag{6.34}$$

This is the condition for stability in such a multidimensional situation, with the competition distributed in a completely interdependent and symmetrical way. In distinction from the earlier one-dimensional multispecies results, this gives a limit to niche overlap which is directly proportional to the environmental variance (as for the pure two-species case in one dimension).

### General Case

In nature, the many resource dimensions that define real niches are unlikely to conform either to the limit of perfect orthogonality or to the opposite limit of perfect interdependence (for further discussion, see the splendid paper by Whittaker, 1969). Even so, the above considerations serve to "bracket" reality, suggesting a limit to species packing and niche overlap. On the one hand, this limit approaches rough independence of the degree of environmental fluctuation ($d/w \sim 1$) as the resource dimensions become independent, and, on the other hand, approaches rough proportionality to the environmental variance ($d/w \sim \sigma$) as the competition is intricately mixed among the resource dimensions.

This discussion has unavoidably been rather imprecise. Its aim was to indicate that the results for the idealized one-dimensional niche can have relevance to the real world, with its multidimensional niches.

## COMPARISON WITH REAL ECOSYSTEMS

Before attempting a confrontation between the natural world and the above theory, we must keep in mind two serious reservations.

First, as just noted, if there is only one relevant resource dimension, or several orthogonal ones, the predictions of the theory are clear-cut, namely a limit at $d/w \sim 1$, roughly independent of the environmental noise level. This facilitates comparison with field data. Conversely, if the multidimensional niche approximates the interdependent simplex, it is difficult to assign meaning to $d$ and $w$, and moreover the limit to niche overlap is a strong function of environmental variance. Comparison with experiment is not to be looked for here.

Second, the theory is restricted to communities whose constituent species' niches are sorted out by competition within a single trophic level. While such an assumption may be relevant for species near the top of the trophic tree, there will be many cases where this single level competition is embedded in the midst of a complex trophic structure. Although intrinsic stability with respect to competition within one trophic level often tends to be associated with overall web stability, no simple generalization is possible: this question was discussed more fully in Chapter 3 (pp. 58–62). In short, interspecific competition can often be the major factor determining niche overlap, but there are many circumstances where other factors, such as predation, are of dominant importance. This point has been stressed by Connell (private communication), and, as one among many examples, Janzen (1970) has suggested explicit mechanisms whereby tropical plant diversity is enhanced by herbivorous predation, particularly seed predation.

Indeed, an interesting future extension of the present theoretical model will be to add a higher trophic level of predation, and to explore the possible consequences.

These two considerations restrict the circumstances where we can look to compare the theory directly with the real world. As previously remarked, some communities may be singled out as approximately one-dimensional. Many bird communities may be meaningfully thought of as competing in the three roughly orthogonal dimensions of vertical habitat, horizontal habitat, and food size. The same may tend to be true of other higher vertebrate communities, for example the Caribbean lizards so thoroughly studied by Schoener (1973). Moreover, for such communities near the top of the trophic ladder, competition is liable to be the predominating influence in species packing. It is among such communities that we look to test the theory. On the other hand many insect communities, and many plant communities, have complex multidimensional niche structure involving, inter alia, a variety of chemical dimensions. In addition, predation and other interactions beyond simple competition are pervasive in these invertebrate and plant communities. Such situations are best avoided for the time being.

(These remarks prompt the observation that people whose field work involves creatures such as birds—where censuses can show remarkable uniformity from year to year—are apt to look more kindly on simplistic theoretical forays of the sort in this chapter than are people whose field work involves, say, rose thrips. In turning to the underlying questions, one should try to dissociate oneself from the prejudices induced by either extreme experience in the field.)

One of the first observations suggesting the existence of a theory such as presented here was Hutchinson's (1959)

166

remark that in a variety of circumstances, including both vertebrate and invertebrate forms, character displacement among sympatric species leads to sequences in which each species is roughly twice as massive as the next; i.e. linear dimensions as measured by bills or mandibles in the ratio 1.2 to 1.4. Interpreted in the light of Hespenheide's (1971) work, as mentioned below, this comes to $d/w \sim 1$. The present state of the evidence is comprehensively reviewed by MacArthur (1972), who concludes that these quantitative data point to there being a limiting value to niche overlap in the natural world, corresponding to $d/w$ in the range 1 to 2.

Also pertinent is Simpson's (1964) review of the factors making for latitudinal and altitudinal species diversity gradients among North American mammals, which concludes that degree of niche overlap is not an important contributing factor. That is, it concludes that the degree of niche overlap is not significantly dependent on the magnitude of the environmental variability.

The work which seems to come closest to our one-dimensional model is that of Terborgh, Diamond, and Beaver on various guilds of birds in an assortment of habitats which have varying degrees of environmental stability. Even so, such comparisons with the theory are necessarily approximate, partly because our competition coefficients come ultimately from utilization functions which are not just percentage of time or of diet, but rather have weighting terms for resource renewal (MacArthur, 1970, 1971b, 1972): all available information from nature contains unweighted utilizations.

Terborgh (1972) has shown that four species of tropical antbird, segregating by foraging height, have mean heights separated by one standard deviation, that is $d/w \simeq 1$; see Figure 6.7. MacArthur's (1971b) analysis of Storer's (1966) data on the food weight distribution for three congeneric

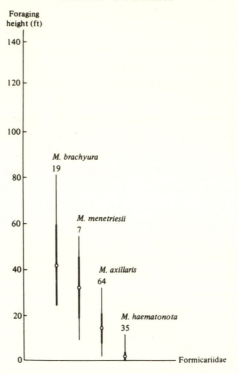

FIGURE 6.7. Foraging height relations within a set of congeneric antbirds (Formicariidae) of the genus *Myrmotherula*. The congeners seem to separate on this single niche dimension. The thick bars denote one standard deviation ($w$), and narrow bars the entire range of observations; above the bars are given the number of observations of each species. (Provided by J. Terborgh, in Mac-Arthur, 1972, p. 43.)

species of hawks also leads to $d/w \simeq 1$. Diamond's (1972) extensive data on weights of tropical bird congeners which sort out largely (but not wholly, so that $d/w$ should be smaller than our one-dimensional theory predicts) by size differences leads to weight ratios around 1.6 to 2.3. Using Hespenheide's analysis (1971) of the relation between weight ratio and $\alpha$ (see also MacArthur, 1972, Ch. 3), Diamond's results become $\alpha \simeq 0.8$ to 0.9, i.e. $d/w \simeq 0.6$

to 1.0. In the Sierra Nevada, Beaver (1972) has shown that species packing in a brushland bird community appears equal to that in a forest foliage gleaning guild, although the microenvironment is thought to be significantly more unvarying in the forest. A pair of salamanders, mentioned as interesting by Hutchinson (1957) because of their similar size (mean length, 55 cm versus 51 cm) and apparently similar ecology, turn out to take diet items differing by a size factor of roughly two (mean of 8.4 items in the stomach of one versus 20.8 items in the slightly smaller stomach of the other: Dumas 1956), again corresponding to $d/w \sim 1$. Nor is one confined to living beasts. The largest soaring bird of prey yet known to have lived is the North American *Teratornis incredibilis*, with a wingspread of around 16 feet; its congener, *T. merriami*, which likewise went extinct a few thousand years ago, had a 12-foot wingspan, again conforming to Hutchinson's weight ratio of around 2, and a deduced $d/w \sim 1$.

More atypical, since they come from the realm of insects and plants rather than birds and other higher vertebrates, are Heatwole's and Davis's (1965) work on three sympatric wasp species, whose ecology seems identical except for their ovipositor lengths, where on this one resource dimension we find $d/w \sim 1$, and Whittaker's (1969) sequence of oak species in the Santa Catalina Mountains of Arizona. Whittaker's results are shown in Figure 6.8 (which is reminiscent of the schematic Figure 6.1), where we see that the two black oaks of the same subgenus have $d/w$ slightly in excess of unity, as do the three white oaks of the same subgenus, while the constellation of all five congeneric oaks together have $d/w$ just less than unity, as does the total system with four other broadleaf tree species added at the ends of the resource (altitude) spectrum. Similar results have been shown by Curtis (1959, p. 94) for Wisconsin oaks.

169

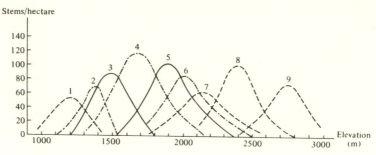

FIGURE 6.8. Distribution of oaks and other broadleaf trees along the altitude gradient in the Santa Catalina Mountains, Arizona. Species shown with solid lines (3, 5) are evergreen oaks of the black oak subgenus *Erythrobalanus*, those with dot-and-dash lines (2, 4, 6) are evergreen oaks of the white oak subgenus *Lepidobalanus,* and those with broken lines (1, 7, 8, 9) are other broadleaf tree species. (After Whittaker, 1969.)

In the foregoing, no distinction has been made between habitat dimensions (such as altitude in Figure 6.8) and other niche dimensions. Thus Grant's (1966) observation of three congeneric pairs of birds, one pair with large bill-length difference but little habitat differentiation, one pair with small bill-length difference and strong habitat differentiation, and the third pair with an intermediate amount of bill-length difference and habitat overlap, are for our purposes three essentially equivalent examples of the rule that $d/w \sim 1$. This is not to deny that a more detailed discussion may care to distinguish various kinds of niche dimensions (e.g. Whittaker, 1969).

In brief, the basic conclusion which emerges in a non-obvious but robust way from the mathematical model, namely that there is a limit to niche overlap in the natural world and that this limit is not significantly dependent on the degree of environmental fluctuation (unless it be severe, as in the arctic), seems to provide a conceptual framework within which a large amount of empirical data may be unified.

## SUMMARY

The evidence for, and conjectures about, the limits to species packing have been graphically put by Whittaker (1969). Likening the competing species to dancers, and the total niche hypervolume to their $n$-dimensional dance floor, he observes: "As additional dancers enter the floor, manoeuvres make space for them, with reduction of the dance areas of the remaining couples. There may come a time, however, when a part of the floor becomes so crowded that the rate at which new dancers enter is equaled by the rate of departure of couples discouraged or crowded off the floor."

As a step toward adding mathematical substance to this graceful image, we study the relation between niche overlap and environmental variability for a class of simple model biological communities in which several species compete on a one-dimensional continuum of resources, e.g. food size. In a strictly unvarying (deterministic) environment, there is in general no limit to the degree of overlap, no limit to the number of dancers. However, in a fluctuating (stochastic) environment, it is found that the average food sizes for species adjacent on the resource spectrum must differ by an amount roughly equal to the standard deviation in the food size taken by either individual species. This limit to species packing has a very weak (logarithmic) dependence on the degree of environmental variance.

This mathematical result is very robust, as is shown by considering, inter alia, a wide range of resource spectrum shapes, and a variety of shapes for the functions describing how the species utilize the resource. The general conclusions seem in harmony with such relevant field data as are available.

171

# CHAPTER SEVEN
# Speculations

In this self-indulgent final chapter, I return to the broad general theme of diversity, complexity and stability in natural ecosystems.

In the very special biological communities of Chapter 6, with only one trophic level, it is clear that greater complexity in the form of more species, more closely packed, makes for less stability. On the other hand, closer species packing (more "Fourier components") tends to give a better fit to the resource spectrum, and a more efficient and total use of the available environmental resources. Thus, if it were not for community stability considerations, evolution would tend to drive toward greater and greater niche overlap. In a perfectly stable deterministic environment, arbitrarily close packing and rich speciation is possible, and in the real world to a certain limited extent the greater the environmental steadiness, the closer the packing, and the richer the consequent assembly of species. Insofar as Chapter 6 adds a piece to the complexity-stability puzzle, it is that complexity is a fragile thing, permitted in this instance by environmental steadiness. This is quite the opposite of the conventional "complexity begets stability" wisdom.

Real biological communities usually comprise not one but many trophic levels, often of labyrinthine complication.

The array of model ecosystems in Chapter 3 indicates that, as a mathematical generality, increased complexity makes for diminished community stability. Even in the natural world — which is no general system — it is not true that population stability is uniformly associated with trophic complexity and faunal and floral diversity. Elton's six arguments for the complexity-stability thesis do not stand up well to close examination: mathematical models and laboratory experiments with simple systems may indeed be unstable, but the analogous multispecies systems are characteristically even less stable; the collapse of island ecosystems measures their vulnerability to novel perturbations rather than their natural stability; while the comparison of stable rain forest versus unstable agrarian monoculture, or orchards, is a comparison of natural versus artificial systems, not just of complex versus simple ones. The U.S. east coast marsh grass *Spartina alterniflora* or the U.K. estuarine *S. townsendii* are stable natural monocultures whose large expanses can well be mistaken by the untutored for man's handiwork. Tischler (1955, 1972) has emphasized other examples, particularly in littoral systems, of productive environments where the stable natural system is essentially a monoculture. What sets man-made monocultures apart is not so much their simplicity as their "unnaturalness," the community's lack of any significant evolutionary pedigree.

Natural ecosystems, whether structurally complex or simple, are the product of a long history of coevolution of their constituent plants and animals. It is at least plausible that such intricate evolutionary processes have, in effect, sought out those relatively tiny and mathematically atypical regions of parameter space which endow the system with long-term stability.

This suggestion, based in part on the properties of general mathematical models, is interesting. It implies that complex and stable natural systems are likely to be fragile, tending to crumple and simplify when confronted with disturbances beyond their normal experience (that is, tending to instability when carried out of their small and particular stable region of parameter hyperspace). This seems consonant with the facts. Curtis (1956) documents the observation that in plant communities the first to go, under the impact of man, are the "upper middle class" plants, which "make up the most advanced communities of a given region from the standpoints of degree of integration, stability, complexity, and efficiency of energy utilization" (p. 734). The view has been put that tropical rain forests come under the heading of Nonrenewable Resources, unable to regenerate after large-scale disturbance (Gómez-Pompa, Vasquez-Yanes, and Guevara, 1972, and references therein).

In short, there is no comfortable theorem assuring that increasing diversity and complexity beget enhanced community stability; rather, as a mathematical generality, the opposite is true. The task, therefore, is to elucidate the devious strategies which make for stability in enduring natural systems. There will be no one simple answer to these questions. Some of the current ideas reviewed in Chapters 3 and 4 may be among the first steps along the road to understanding. Although these models are admittedly grossly oversimplified, it may be remembered that the crudities of Bohr's hydrogen atom were a useful start toward understanding the complications of molecular spectra.

Until such time as we better understand the principles which govern natural associations of plants and animals, we would do well to preserve large chunks of pristine ecosystems. They are unique laboratories. Quite apart from valid ethical and aesthetic considerations, there are pragmatic reasons why we should query the increasingly universal replacement of natural ecosystems, with their long evolutionary history, by agroecosystems, which are usually intrinsically unstable.

## THE LATITUDINAL DIVERSITY GRADIENT

Some facets of the above issues are reflected in the widely recognized decrease in species diversity going from the tropics to the poles.

Most of the ideas advanced to account for this latitudinal diversity gradient (Darlington, 1957, 1965; Miller, 1967; Klopfer, 1962; Pianka, 1966; MacArthur, 1972, Ch. 8) may be subsumed under three general headings. The three factors which, in principle, determine the number of species present in a given environment are:

(i) The total (multidimensional) niche volume, $V$, available in the particular environment;

(ii) The effective niche volume, $v$, for a typical individual species;

(iii) Historical aspects, relating to how much time there has been for evolution and adaptation in the given environment.

Succinctly, the potential number of species is the total niche volume divided by the effective niche volume per species, $V/v$, and this potential will be realized if enough time is available:

$$\text{Potential number of species} \sim V/v. \qquad (7.1)$$

We now briefly discuss these three factors. A similar formal framework is developed by MacArthur (1972, Ch. 7).

(i) The total niche hypervolume, $V$, is greater in the tropics, which tend to be more productive and less seasonal. They are also more floristically complex, both in stratification and diversity, which makes for more and wider resource dimensions for the fauna (this remark, of course, begs the question as to plant diversity). Moreover

175

the above linear factorization is oversimplified in that increasingly rich speciation and specialization can themselves create new resource dimensions, thus expanding the total niche volume. Overall, there are strong arguments for a latitudinal gradient in $V$.

(ii) The effective niche volume per species, $v$, depends basically on the niche-to-niche spacing $d$ (cf. Chapter 6 and Figure 6.1). This may be usefully reexpressed as the product of the intrinsic niche width, $w$, and the degree of niche overlap, $d/w$,

$$d = (w)(d/w). \tag{7.2}$$

The intrinsic niche width, $w$, will be smaller for specialist species than for generalists. As discussed in detail by MacArthur and Pianka (1966), Schoener (1969), and Roughgarden (1972), the "economics of consumer choice" (MacArthur, 1972, Ch. 3) suggest that a productive, and above all a steady or predictable, environment will favor greater specialization. This clearly is the case in nature, and makes for greater specialization in the tropics. The tendency is reinforced by the fact that relatively benign and productive environments may support a larger biomass, and consequently a larger gene pool, which assists the underlying genetic mechanisms whereby specialization evolves (Darlington, 1957, Ch. 9; Connell and Orians, 1964).

The other relevant quantity in equation (7.2), the niche overlap $d/w$, was seen in Chapter 6 to attain an effective limiting value of roughly unity: $d/w \sim 1$ (cf. Figure 6.3). This limit is usually only weakly dependent on the degree of environmental fluctuation, unless it be very severe, whereupon the limit to $d/w$ is significantly in excess of unity.

Thus Chapter 6 constitutes an important piece of the latitudinal diversity gradient picture. It suggests that niche-to-niche spacing $d$, and thence per species niche volume $v$, essentially depends only on the intrinsic niche width $w$ (cf. equation (7.2)). The above arguments as to large versus small $w$ then point to $v$ being smaller in the tropics. Insofar as degree of niche overlap can depend on the environmental variability, the effect gives smaller $d/w$, and thence yet smaller $v$, in the tropics

(iii) The preceding discussion pertains to the equilibrium species' density, $V/v$, attained after the system has shaken itself down into some broadly enduring configuration. This assumes the environment in question has remained roughly unchanged over times significant on an evolutionary scale. When this is not so, the vagaries of biogeographic history can lie between the potentiality and the actuality. Darlington (1957, 1965) has made this point. As a matter of history, there has been more time for speciation in the tropics than in most other places, which could contribute to the latitudinal diversity gradient. The role played by time in the evolution of intricately diverse systems is underlined by the comparatively rich speciation in the "old" lakes, Malawi and Baikal, at either end of the Great Rift, or by Southwood's (1961) evidence (culled from several countries but mainly from Britain) that "the number of insect species associated with a tree is a reflection of the cumulative abundance of that tree in the particular country throughout recent geological history, i.e. the Quaternary period," which means that historically dominant native trees will have most insect species, and recently introduced ones fewest.

Another way history can enter is in determining into which of a number of alternative equilibrium configurations a system will settle down. This, however, probably has little bearing, as the various equilibrium values of $V/v$ are likely to be roughly equal.

177

## HOW MANY SPECIES?

The discussion so far has been of contemporary patterns of species diversity, of the most recent frame in the movie of biological history. The fossil record provides a rough means of running this movie backward, of comparing the present with the past.

Figure 7.1 shows the total fossil record of aquatic animals, in terms of the number of orders known for each readily fossilizable phylum in each of the conventional geological epochs. These geological periods are indicated sequentially, without being scaled to their actual duration. For the first ten periods (Cambrian, Ordovician, Silurian, Devonian, Carboniferous, Permian, Triassic, Jurassic, Cretaceous, Tertiary) this is sensible enough, as each lasted

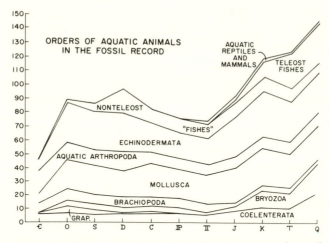

FIGURE 7.1. The record of readily fossilizable groups of aquatic animals. Numbers of orders known within each named phylum in each period are plotted cumulatively on the y-axis. The geological periods, indicated by their conventional abbreviations (see the text) are placed sequentially on the x-axis but are not scaled to absolute time. (Ater Simpson, 1969.)

about 60 million years, adding up to 600 million years; some epochs (e.g. Cambrian) are a little longer, others (e.g. Triassic, Jurassic) a little shorter. The final period, the Quaternary, is a brief 1 million years, and the final upswing is therefore on a quite distorted time scale. Indeed the upswing in the Quaternary in Figure 7.1 is due wholly to soft-bodied orders of bryozoans and coelenterates which are likely to have arisen earlier, without leaving a fossil signature: this upswing should be ignored. Raup (1972) has discussed in detail some of the biases which creep into the fossil record, especially at lower taxonomic levels such as species or genera; he presses the view that even at the level of species the increase in marine diversity since the Paleozoic (Cambrian through Permian) is more apparent than real. These sampling difficulties are, however, largely avoided by working at the higher taxonomic levels of orders (Figures 7.1 and 7.2) or superfamilies (Figure 7.3).

The idea behind Figure 7.1 is that increase in the number of members in the higher taxonomic categories (here, specifically, orders) should provide a rough indication of increase in the ecological complexity and diversity of communities. The evolution of new orders usually corresponds to distinct adaptive types, which radiate into specific niches at lower taxonomic levels. Although there are, of course, many complications and qualifications, Simpson (1969, p. 167) has written "it is reasonable to believe that in following the numbers of orders in the fossil record we are indeed following the approximate overall course of ecological complication and diversification even though in a necessarily loose way."

It is seen that the number of orders rose rather steadily throughout the Cambrian into the Ordovician. Although the relative contributions from the various phyla change, the total number of orders remains roughly constant at 80–100 from the Ordovician through the Silurian, Devo-

nian, and Carboniferous to the Permian. The dip around
the Triassic is a well-established feature of the fossil
record. It reflects the extinctions and faunal turnover
which occurred in this epoch, and may be ascribed to
geological and climatic factors (e.g. uplift of continents,
withdrawal of seas, deteriorating climate). The subsequent
recovery in the Cretaceous and Tertiary may then be a
return to the ecological norm of the earlier part of the
record. It must be remembered that, at all times in Figure
7.1, in most phyla major replacements of lower taxonomic
groups is going on, but without notable changes in their
numbers.

Thus, after the initial build-up in the Cambrian, the
subsequent 500 million years exhibits a remarkable steadiness in the total number of orders of animals in the sea.
In Simpson's (1969, p. 168) words, "The Cretaceous-
Tertiary return to approximately the Ordovician-Devonian
level is a sort of stabilization of ordinal numbers, suggesting
that there is an approximate number of orders of aquatic
animals that is 'right' in the sense of filling the usually
available ecological situations."

Figure 7.2 is the analogue of Figure 7.1, this time for
land-dwelling and amphibious animals.

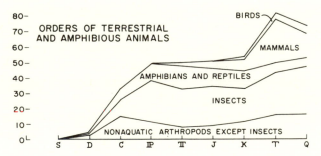

FIGURE 7.2. The record of readily fossilizable groups of land-
dwelling and amphibious animals. The construction and conven-
tions as to x- and y-axes are as in Figure 7.1. (After Simpson, 1969.)

We note that, whereas the seas were essentially ecologically "saturated" with orders by the late Ordovician, it was not until the Silurian that land communities began to arise. These communities developed more definitely in the Devonian, and seem to have reached a sort of sustained plateau of numbers of orders (around 50) by the Permian. The rise in the multiplicity of mammalian orders after the Cretaceous is in large part due to the fact that a number of different orders came to play essentially the same ecological role in separate continents: earlier ages contained fewer physiographically distinct land areas. This point has been documented in some detail by Kurten (1969). Once the same broadly defined ecological niche is filled by different orders in different continents, counting total numbers of orders gives an unreliable measure of the number of distinct terrestrial niches. This complication is less of a feature in Figure 7.1 because the great majority of aquatic orders are widespread, presumably since geographical barriers are less common in the oceans. Again it is to be remembered that the Quarternary period is only 1 million years, and thus out of scale on Figure 7.2. The marked Quaternary downturn in mammalian diversity may be associated with the rise of man (Martin, 1966; Ho, 1967).

As mentioned earlier, many complications are hidden by looking at groups at the taxonomic level of orders. If we look at the insect phylum at the level of species, genera, or even families, there are great changes with the spread of flowering plants; but these changes do not show up significantly in changing the number of orders in the phylum. This may be deemed a virtue, rather than a vice, of looking at things at this high a taxonomic level.

Viewed in broad qualitative terms, the preceding results support the idea that under given circumstances there is an approximately fixed number of species which fill an environment. Once filling has occurred, which may take a long time (for example Cambrian-Ordovician for marine environments), the number of taxa tends to stabilize over long time scales. This is not to deny that essentially the same ecological roles may be, and in fact usually are, played by quite different taxa in the course of geological time. It is the number of distinct "ways of life," the number of distinct niches, that is the roughly conserved quantity. Examples of this phenomena abound in the phyla of Figures 7.1 and 7.2, and have been documented by Simpson (1953, 1965, 1969), Romer (1966), and others. We mention only two.

Figure 7.3 provides a schematic summary of the evolutionary history of the ammonites, an extinct group of mollusks occupying one broad adaptive zone, divided into a large number of narrow generic zones and particular niches. Three times, at the end of the Devonian, the Permian, and the Triassic, this order was greatly reduced in

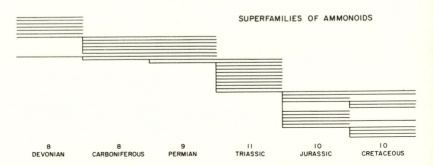

SUPERFAMILIES OF AMMONOIDS

| 8 | 8 | 9 | 11 | 10 | 10 |
| DEVONIAN | CARBONIFEROUS | PERMIAN | TRIASSIC | JURASSIC | CRETACEOUS |

FIGURE 7.3. Known superfamilies of ammonoidae. Each superfamily is indicated by a horizontal line in the geological period when it occurred (times of appearance or disappearance within periods are not indicated). Vertical lines tie descendent superfamilies (below) to their supposed ancestry (above). As before, the *x*-axis simply labels periods, and is not to scale. Notice the remarkable constancy in the number of superfamilies. (After Simpson, 1969, following Moore, 1967.)

182

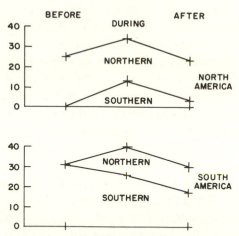

FIGURE 7.4. Numbers of families of land mammals involved in faunal interchange between North and South America in the late-tertiary/early-quaternary epoch: the families are classified according to whether they were southern or northern before interchange. The y-axis indicates total numbers of families. The x-axis is not scaled to absolute time, but indicates the sequence of three phases: before (several million years ago), during (roughly one million years ago), and after the interchange. (After Simpson, 1969.)

variety (presumably by the climatic and other environmental shocks which mark the interfaces between the conventional geological epochs). Each time, a small number of taxa, either one or two as indicated in the figure, did survive, and in each case the survivors subsequently diversified until approximately the same level of taxonomic (hence probably of ecological) diversity was regained. Thus the history of the order is characterized by a sequence of discontinuities, with a return to relative stability in terms of ecological diversity after each such discontinuity. Finally, in the late Cretaceous, an otherwise similar discontinuity left no survivors; it may be conjectured that

183

another order or orders subsequently filled these niches.

A different example, on a much shorter time scale, is shown in Figure 7.4. At the beginning of the Quaternary or Pleistocene period there was no land connection between North and South America, and there were no families of terrestrial mammals in common between the two continents. Sometime in the mid-Pleistocene the present land connection was formed, and there ensued a great faunal interchange between the two continents. During the interchange, the total number of families can be seen to have risen markedly in both continents, but subsequent extinctions eventually caused the numbers of families on both continents to drop back to approximately the numerical level that existed before the faunal mixing. In Simpson's (1969, p. 174) words, "This strongly suggests, or at the very least is consistent with, the idea that each continent was ecologically full of land mammals before interchange and that the number of about 25 families in North, about 30 in South America represents ecological saturation."

Overall, if one looks at figures such as 7.1 to 7.4 in terms of the biological details of the individual taxa, one sees ceaseless change and flux. But if one looks at them as providing information about the number of broadly defined niches, they give striking evidence of approximate constancy over geological time spans.

This evidence as to the grand strategy of community evolution admits of many interpretations.

The extreme cynic may suggest that taxonomists are apt to end up with the same constant, predetermined number of taxa, no matter what data they work with.

Alternatively, it may be argued that there is no limit to speciation and niche divisibility, no limit to the number of taxonomically distinct groups in a given environment, but that global environmental discontinuities on a geological time scale keep interrupting this process of ever finer and finer differentiation, thus creating the illusion of constancy in the number of taxa. This view cannot be altogether ruled out, although it does require such macroscopic geographic and climatic changes to occur at roughly uniform intervals, and with a time scale short compared to that for adaptive evolutionary processes.

While conceding that such ideas may play minor parts in the total story, I favor the view that the apparent constancy evidenced by the data mirrors a real constancy. The gist of this opinion has been well put by Darlington (1957): "Throughout the recorded history of vertebrates, whenever the record is good enough, the world as a whole and each main part of it has been inhabited by a vertebrate fauna which has been reasonably constant in size and adaptive structure. Neither the world nor any main part of it has been overfull of animals in one epoch and empty in the next, and no great ecological roles have been long unfilled. There have always been (except perhaps for very short periods of time) herbivores and carnivores, large and small forms, and a variety of different minor adaptations, all in reasonable proportion to each other. Existing faunas show the same balance. Every continent has a fauna reasonably proportionate to its area and climate, and each main fauna has a reasonable proportion of herbivores, carnivores, etc. This cannot be due to chance" (p. 553). "Something holds the size and composition of faunas in all parts of the world and at all times within certain limits, in spite of continual changes and successions in separate phylogenetic groups" (p. 620).

If this argument is accepted, the job of estimating roughly how many broadly defined niches a given environment will have—and thus of crudely estimating the ordinal numbers on the y-axis in Figures 7.1 to 7.4—becomes a fascinating task for the population biology of the future.

185

Of course, this is mere speculation, and brash speculation at that. Moreover, even if such a task be fulfilled, the legitimate interest of most biologists will still be the differences between ecosystems, the splendid and variegated details that make each natural community unique, rather than colorless order-of-magnitude generalities as to the number of constituent species. In a like fashion, for most people the interesting thing about rocks is the way differences in crystal structure, mode of deposition, and geological history give rise to a kaleidoscopic variety of forms; but it is also nice to understand why all rocks have roughly (to within a factor of ten) the same density, and to be able to provide a theoretical estimate of that density.

Provided that enough time for ecological "saturation" has elapsed (the factor (iii) of p. 175), the task of estimating the number of species or other taxonomic groups in a given environment comes down to estimating the $V/v$ of equation (7.1). At the present stage of knowledge, very little can sensibly be said about estimating either the total niche volume, $V$, or the effective individual niche volume, $v$. The current explorations as to the principles underlying trophic web structures, and particularly the work on the limits to niche overlap and similarity, are a few among many initial elements required in this hubristic view to the future.

# Appendices

This is the Cybernetics and Stuff
That covered Chaotic Confusion and Bluff
That hung on the Turn of a Plausible Phrase
And Thickened the Erudite Verbal Haze
Cloaking Constant K
That saved the Summary
Based on the Mummery
Hiding the Flaw
That lay in the theory Jack built.

From *The Space Child's Mother Goose*,
Winsor and Parry, 1958

187

# APPENDIX I

For the benefit of certain classes of reader, this appendix puts some flesh on the bare bones of the formalism in Chapter 2. We illustrate how the community matrix is constructed, and how the neighborhood stability properties are found from it, for two particular examples. These are (i) the familiar Lotka-Volterra one-predator–one-prey model, and (ii) the more complicated predator-prey system of equation (4.5).

For more thorough exposition of these general ideas, see for example Maynard Smith (1968), Rosen (1970), or Barnett and Storey (1970).

*Example (i)*. The Lotka-Volterra equations (3.1) are one particular case of the general formal equations (2.8) of Chapter 2. Writing the prey population as $N_1$ and the predator population as $N_2$, these Lotka-Volterra equations correspond to equations (2.8) with $m = 2$ and

$$F_1(N_1, N_2) = N_1[a - \alpha N_2] \tag{A.1a}$$

$$F_2(N_1, N_2) = N_2[-b + \beta N_1]. \tag{A.1b}$$

The equilibrium populations $N_1^*$ and $N_2^*$ are the non-zero solutions obtained by putting $F_1 = F_2 = 0$ (that is, equation (2.9)), which clearly gives

$$N_1^* = b/\beta; \; N_2^* = a/\alpha. \tag{A.2}$$

The elements $a_{ij}$ of the community matrix can now be calculated from the prescription (2.13). We begin by working out all the partial derivatives $\partial F_i/\partial N_j$:

$$\partial F_1/\partial N_1 = a - \alpha N_2$$

$$\partial F_1/\partial N_2 = -\alpha N_1$$

$$\partial F_2/\partial N_1 = \beta N_2$$

$$\partial F_2/\partial N_2 = -b + \beta N_1.$$

Evaluated at the equilibrium point, equation (A.2), these partial derivatives give

$$a_{11} = 0, \ a_{12} = -\alpha b/\beta,$$

$$a_{21} = \beta a/\alpha, \ a_{22} = 0, \tag{A.3}$$

giving the community matrix (3.2).

The eigenvalues of the matrix follow from the determinantal equation (2.17), which here takes the form

$$\det \begin{vmatrix} -\lambda & -\alpha b/\beta \\ \beta a/\alpha & -\lambda \end{vmatrix} = 0.$$

That is

$$\lambda^2 + ab = 0. \tag{A.4}$$

Thus the eigenvalues are the pair of purely imaginary numbers $\pm i\omega$, where for notational convenience we introduce $\omega = \sqrt{ab}$.

The perturbations to prey and predator populations are linear combinations of the factors $\exp(\lambda_1 t)$ and $\exp(\lambda_2 t)$, with coefficients depending on the initial disturbance (cf. equation (2.14)). Here this means we have linear combinations of the purely oscillatory factors $e^{i\omega t}$ and $e^{-i\omega t}$, which is to say linear combinations of $\cos(\omega t)$ and $\sin(\omega t)$. That is, the stability is neutral, with perturbations leading to undamped pure oscillations, of frequency $\omega$ or period $2\pi/\omega$.

For this Lotka-Volterra model, a Lyapunov function (see p. 15) can be constructed, so that this neighborhood analysis validly describes the global stability character.

*Example (ii).* A similar, but more elaborate, example is provided by the one-predator–one-prey system of equation (4.5). Again writing the prey and predator populations as

$N_1$ and $N_2$, respectively, we have a particular realization of the formal equations (2.8) with $m = 2$ and

$$F_1(N_1, N_2) = rN_1 \left[ 1 - \frac{N_1}{K} - \frac{(k/r)N_2}{N_1 + D} \right] \qquad \text{(A.5a)}$$

$$F_2(N_1, N_2) = sN_2 \left[ 1 - N_2/(\gamma N_1) \right]. \qquad \text{(A.5b)}$$

The equilibrium populations $N_1^*$ and $N_2^*$ are obtained from equation (2.9), $F_1 = F_2 = 0$:

$$N_2^* = \gamma N_1^* \qquad \text{(A.6)}$$

$$1 - \frac{N_1^*}{K} - \frac{(k\gamma/r)N_1^*}{N_1^* + D} = 0. \qquad \text{(A.7)}$$

It is convenient to define the two quantities

$$\alpha \equiv k\gamma/r \qquad \text{(A.8)}$$

$$\beta \equiv D/K. \qquad \text{(A.9)}$$

The solution of the quadratic equation (A.7) for $N_1^*$ may then be written

$$N_1^* = D(1 - \alpha - \beta + R)/(2\beta), \qquad \text{(A.10)}$$

with the definition

$$R \equiv [(1 - \alpha - \beta)^2 + 4\beta]^{1/2}. \qquad \text{(A.11)}$$

Thus having determined the possible equilibrium populations, we proceed to the neighborhood stability analysis.

To construct the community matrix, we evaluate all the partial derivatives $(\partial F_i/\partial N_j)^*$ at the point $N_1^*, N_2^*$. This gives

$$a_{11} = \left( \frac{\partial F_1}{\partial N_1} \right)^* = rN_1^* \left[ -\frac{1}{K} + \frac{(k/r)N_2^*}{(N_1^* + D)^2} \right]$$

$$a_{12} = \left( \frac{\partial F_1}{\partial N_2} \right)^* = -\frac{kN_1^*}{N_1^* + D}$$

$$a_{21} = \left( \frac{\partial F_2}{\partial N_1} \right)^* = \frac{s(N_2^*)^2}{\gamma(N_1^*)^2} = \gamma s$$

$$a_{22} = \left(\frac{\partial F_2}{\partial N_2}\right) = -\frac{sN_2^*}{\gamma N_1^*} = -s.$$

The determinantal equation (2.17) for the eigenvalues $\gamma$ reduces to the quadratic equation

$$\lambda^2 + a\lambda + b = 0, \qquad (A.12)$$

with

$$a = -(a_{11} + a_{22})$$

$$b = a_{11}a_{22} - a_{12}a_{21}.$$

The necessary and sufficient condition for the system to possess neighborhood stability is that both eigenvalues $\lambda$ have negative real parts. This (see Appendix II) requires both the coefficients $a$ and $b$ in (A.12) to be positive. The condition $b > 0$ is automatically fulfilled:

$$b = rsN_1^* \left[ +\frac{1}{K} - \frac{\alpha N_1^*}{(N_1^* + D)^2} + \frac{\alpha}{(N_1^* + D)} \right] > 0.$$

The other condition, $a > 0$, clearly requires

$$s - rN_1^* \left[ -\frac{1}{K} + \frac{\alpha N_1^*}{(N_1^* + D)^2} \right] > 0.$$

After some algebraic manipulation, using equation (A.10) for $N_1^*$ along with the definitions (A.8), (A.9) and (A.11), this can be written as the stability criterion

$$\frac{s}{r} > \frac{2(\alpha - R)}{1 + \alpha + \beta + R}. \qquad (A.13)$$

In particular, for given $\alpha$ the largest value the right-hand side can attain is $1/\alpha$ if $\alpha > 1$ or $(2\alpha - 1)$ if $\alpha < 1$ (this limit being attained as $\beta \to 0$). Values of $s/r$ in excess of these limits consequently always imply stability; notice that $\alpha < \frac{1}{2}$ is stable for all $s/r$.

The stability condition (A.13) may, or may not, be satisfied, depending on the values of the relevant parameters

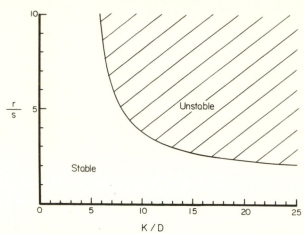

FIGURE A.1. This figure illustrates the stability condition (A.13), in terms of the parameter ratios $r/s$ and $K/D$ $(=1/\beta)$, for the predator-prey system defined by equation (4.5). In the unhatched region of this parameter space, the equilibrium point is stable; in the hatched region, it is unstable. (The figure is for $ky/r = 1$, but qualitatively similar stability boundaries pertain to other values of this ratio.)

$\alpha$, $\beta$ and $r/s$. Although the above was a neighborhood stability analysis, the nonlinear considerations of Chapter 4 show that for this particular model there is either an equilibrium point which is globally stable, or else a stable limit cycle. The criterion (A.13) therefore divides the region where the predator-prey equations (4.5) have a stable equilibrium point from the region where they exhibit stable limit cycle behavior.

In Figure A.1 we fix $\alpha$ to be unity, and show what the stability criterion (A.13) implies about the parameters $r/s$ and $K/D$. Similar stability boundaries could be shown for other values of $\alpha > \frac{1}{2}$. The figure bears out the general remarks made in Chapter 4, namely that there is a stable equilibrium point provided either one of $K/D$ or $r/s$ is not too large. However if both $K/D$ and $r/s$ are largish (corresponding to relatively weak prey self-limitation, and a prey intrinsic growth rate significantly in excess of the predator one), the equilibrium point is not stable, and limit cycle behavior ensues.

# APPENDIX II

This appendix is a compendium of generally useful results relating to the eigenvalues of various matrices. Some of the results are familiar to most people and are presented without proofs (which are to be found in any matrix algebra text). Other results, for example the eigenvalues of the classes of matrices catalogued under the subheading $(v)$ below, are less well known, and their proofs are sketched.

### $(i)$ Symmetric and Antisymmetric Matrices

The eigenvalues of a symmetric matrix are all real numbers.

The eigenvalues of an antisymmetric matrix are all purely imaginary numbers, and occur as conjugate pairs, $+i\omega$ and $-i\omega$. As a corollary it follows that an antisymmetric matrix of odd order necessarily has one eigenvalue equal to zero; antisymmetric matrices of odd order are singular. This causes difficulties in antisymmetric Lotka-Volterra models with an odd number of species, as mentioned in Chapter 3.

In essence, the proofs consist of showing that for a symmetric matrix the eigenvalues are equal to their complex conjugates ($\lambda^* = \lambda$); for an antisymmetric matrix, $\lambda^* = -\lambda$.

### $(ii)$ $U^{-1}AU$ Versus A

Let $U$ be any nonsingular $m \times m$ matrix, and $U^{-1}$ its inverse. Then if $A$ is any $m \times m$ matrix, the eigenvalues of the matrix $U^{-1}AU$ are identical with those of $A$.

Proof: The eigenvalues $\lambda$ of $U^{-1}AU$ are given by

$$0 = \det |U^{-1}AU - \lambda I|$$
$$\equiv \det |(U^{-1})(A - \lambda I)(U)|.$$

The determinant of a product is the product of the determinants, whence

$$0 = \det |A - \lambda I|.$$

Thus the eigenvalues $\lambda$ are identically those of the matrix $A$.

This result was of direct use in Chapter 5 (following equation (5.31)). Some corollaries are also worth noting.

*Corollary I.* Suppose $V$ is a nonsingular $m \times m$ symmetric matrix, and $A$ a $m \times m$ symmetric [antisymmetric] matrix. In general the matrix $VA$ is not symmetric [antisymmetric]. However it remains true that the eigenvalues of $VA$ are all real [all purely imaginary].

Proof: Consider a symmetric matrix $V^{1/2}$, whose square is $V$, and whose inverse is $V^{-1/2}$. Then we know that $(V^{-1/2})(VA)(V^{1/2}) = V^{1/2}AV^{1/2}$ has the same eigenvalues as $VA$. But if $A$ is symmetric [antisymmetric], so is $V^{1/2}AV^{1/2}$. Consequently the eigenvalues of $VA$ are real [imaginary].

This result was used in Chapter 3, where $V$ was the diagonal matrix of the population values: $(V)_{ij} = N_i \delta_{ij}$.

*Corollary II.* Levins (1968a, pp. 50, 64) defines $p_{ih}$ to be the proportion of resource (or habitat, etc.) in category $h$, among the resources used by species $i$. From these we may construct quantities

$$P_{ij} = \sum_h p_{ih}p_{jh}.$$

Levins defines the competition matrix to have elements $\alpha_{ij}(I) = P_{ij}/P_{ii}$; essentially this is to say $\alpha_{ij} \sim P_{ij}$, with a normalization constant $P_{ii}$. On the other hand, our definition (equation (6.5)) of the competition coefficients in

194

Chapter 6 corresponds to the definition $\alpha_{ij}(II) = P_{ij}/[P_{ii}P_{jj}]^{1/2}$; again $\alpha_{ij} \sim P_{ij}$, but the normalization is now such as to make the $\alpha$ matrix manifestly symmetric.

If we define $D$ to be the diagonal matrix whose elements are $(D)_{ij} = [P_{ii}]^{-1/2}\delta_{ij}$, we see that our $\alpha$ matrix is related to Levins' by

$$\alpha(II) = D^{-1}\alpha(I)D, \tag{A.14}$$

and therefore they have the same eigenvalues. Thus, although the different choices of normalization for the elements of the competition matrix can undoubtedly be confusing, the essential matrix properties — the eigenvalues — are the same in all variations.

### (iii) Necessary Conditions for Stability

In Chapter 2 we saw that the necessary and sufficient conditions for neighborhood stability of a system were that all the eigenvalues of some matrix $A$ lie in the left-hand half of the complex plane.

Quite generally, for a $m \times m$ matrix $A$ with elements $a_{ij}$ the sum, and the product, of the eigenvalues are:

$$\text{sum of eigenvalues} = \sum_{i=1}^{m} a_{ii} \equiv \text{Trace } A, \tag{A.15}$$

$$\text{product of eigenvalues} = \det |A|. \tag{A.16}$$

Thus one necessary (but not sufficient) condition which must be satisfied if all the eigenvalues are to have negative real parts is

$$\text{Trace } A < 0. \tag{A.17}$$

In the special case of Trace $A = 0$, either at least one eigenvalue must lie in the right half plane (unstable system), or all eigenvalues must be purely imaginary (the pathological case of neutral stability). Another such necessary, but not sufficient, condition can be seen to be

$$(-1)^m \det |A| > 0. \tag{A.18}$$

### (iv) Routh-Hurwitz Stability Criterion

If $A$ is a $m \times m$ matrix, the equation (2.17) for the eigenvalues $\lambda$ comes down to an $m$th order polynomial equation

$$\lambda^m + a_1\lambda^{m-1} + a_2\lambda^{m-2} + \cdots + a_m = 0. \qquad (A.19)$$

A formal general expression (the Routh-Hurwitz criterion) can now be written down, giving constraints on the coefficients $a_1$, $a_2$, ..., $a_m$ which are necessary and sufficient to ensure all eigenvalues lie in the left half complex plane.

Rather than explain this abstract expression, which no one in their right mind is going to use on $m > 5$ anyway, we catalogue the explicit Routh-Hurwitz stability conditions for $m = 2$, 3, 4, and 5.

$m = 2$ $\qquad\qquad\qquad a_1 > 0; \; a_2 > 0.$ $\qquad\qquad\qquad$ (A.20)

$m = 3$ $\qquad\qquad\qquad a_1 > 0; \; a_3 > 0; \; a_1a_2 > a_3.$ $\qquad$ (A.21)

$m = 4$ $\qquad\qquad\qquad a_1 > 0; \; a_3 > 0; \; a_4 \gg 0;$

$$a_1a_2a_3 > a_3^2 + a_1^2a_4. \qquad (A.22)$$

$m = 5$ $\qquad\qquad\qquad a_i > 0 [i = 1, 2, 3, 4, 5];$

$$a_1a_2a_3 > a_3^2 + a_1^2a_4;$$

$$(a_1a_4 - a_5)(a_1a_2a_3 - a_3^2 - a_1^2a_4) > a_5(a_1a_2 - a_3)^2 + a_1a_5^2. \qquad (A.23)$$

### (v) Eigenvalues of Some Special Matrices

*Case I.* The $m \times m$ matrix with diagonal elements 1, and all other elements $\alpha$,

$$A = \begin{pmatrix} 1 & \alpha & \alpha & \cdot & \alpha \\ \alpha & 1 & \alpha & \cdot & \cdot \\ \alpha & \alpha & 1 & \cdot & \cdot \\ \cdot & \cdot & \cdot & \cdot & \cdot \\ \alpha & \cdot & \cdot & \cdot & 1 \end{pmatrix} \qquad (A.24)$$

is well known to have eigenvalues (e.g. Levins, 1968a, Ch. 3)

$$\lambda = 1 - \alpha \text{ [with multiplicity } (m - 1)]$$
$$\lambda = 1 + (m - 1)\alpha \text{ [once].}$$
$$(A.25)$$

*Case II.* The $m \times m$ matrix with diagonal elements 1, superdiagonal and subdiagonal elements $\alpha$, and all other elements zero,

$$A = \begin{pmatrix} 1 & \alpha & 0 & 0 & \cdot & 0 \\ \alpha & 1 & \alpha & 0 & \cdot & \cdot \\ 0 & \alpha & 1 & \alpha & \cdot & \cdot \\ 0 & 0 & \alpha & 1 & \cdot & \cdot \\ \cdot & \cdot & \cdot & \cdot & \cdot & \cdot \\ 0 & \cdot & \cdot & \cdot & \cdot & 1 \end{pmatrix} \qquad (A.26)$$

was studied by Laplace. Defining the $m$th order determinant $D(m) = \det |A(m) - \lambda I|$, the eigenvalues can be found from the recursion relation

$$D(m) = (1 - \lambda)D(m - 1) - \alpha^2 D(m - 2).$$

The solution of this difference equation can be written

$$D(m) = \alpha^m \frac{\sin (m + 1)\theta}{\sin \theta}, \qquad (A.27)$$

with the definition $\cos \theta = (1 - \lambda)/(2\alpha)$. It follows that the eigenvalues $\lambda_k$ $[k = 1, 2, \ldots, m]$ are

$$\lambda_k = 1 - 2\alpha \cos \left( \frac{\pi k}{m + 1} \right). \qquad (A.28)$$

*Case III.* Particularly interesting is the general class of $m \times m$ matrices whose rows are cyclic permutations of the first one

197

$$A = \begin{pmatrix} c_0 & c_1 & c_2 & c_3 & \cdot & c_{m-1} \\ c_{m-1} & c_0 & c_1 & c_2 & \cdot & \cdot \\ c_{m-2} & c_{m-1} & c_0 & c_1 & \cdot & \cdot \\ c_{m-3} & c_{m-2} & c_{m-1} & c_0 & \cdot & \cdot \\ \cdot & \cdot & \cdot & \cdot & \cdot & \cdot \\ c_1 & & \cdot & \cdot & \cdot & c_0 \end{pmatrix} \qquad (A.29)$$

The eigenvalues $\lambda_k$ $[k = 0, 1, \ldots, m - 1]$ are given by the expressions

$$\lambda_k = \sum_{\ell=0}^{m-1} c_\ell \exp\left[\frac{2\pi i}{m} k\ell\right]. \qquad (A.30)$$

Proof (Berlin and Kac, 1952): Let $\Delta$ be the matrix

$$\Delta = \begin{pmatrix} 1 & 1 & 1 & \cdot & 1 \\ r_0 & r_1 & r_2 & \cdot & \cdot \\ r_0^2 & r_1^2 & r_2^2 & \cdot & \cdot \\ \cdot & \cdot & \cdot & \cdot & \cdot \\ r_0^{m-1} & \cdot & \cdot & \cdot & r_{m-1}^{m-1} \end{pmatrix}$$

where the $m$ quantities $r_k = \exp[(2\pi i/m)k]$ are the $m$th roots of unity. By using the cyclic properties of the matrix $A$, and the relation $r^m = 1$, the elements of the matrix $A\Delta$ can be seen to be

$$(A\Delta)_{ij} = \left[\sum_{\ell=0}^{m-1} c_\ell r_{j-1}^\ell\right] r_{j-1}^{i-1}.$$

That is, with the above definition (A.30) for $\lambda$,

$$(A\Delta)_{ij} = \lambda_{j-1}(\Delta)_{ij}.$$

And, since the matrix $\Delta$ is nonsingular, these $\lambda$ are manifestly the eigenvalues of $A$. Berlin and Kac also construct the eigenvectors, but we will not pursue this.

If the matrix $A$ is symmetric, so that $c_i = c_{m-i}$ ($i = 1, 2, \ldots$), it is easily seen from (A.30) that the eigenvalues occur in pairs $\lambda_k = \lambda_{m-k}$, with the exception of $\lambda_0$ (and of $\lambda_{m/2}$ if $m$ is even). The expression (A.30) reduces to

$$\lambda_k = \sum_{\ell=0}^{m-1} c_\ell \cos\left(\frac{2\pi k\ell}{m}\right). \qquad (A.31)$$

If we put $c_0 = 1$, and all other $c_i = \alpha$, we recover the familiar result (A.25) for the matrix (A.24).

If, further, the matrix elements steadily decrease as one moves away from the diagonal, $c_0 > c_1 > c_2 > c_3 \ldots$, as is the case in many interesting contexts, the minimum eigenvalue can be seen in the limit $m \gg 1$ to be

$$\lambda_{\min} = c_0 - 2c_1 + 2c_2 - 2c_3 + \cdots \qquad (A.32)$$

The maximum eigenvalue is $\lambda_0$,

$$\lambda_{\max} = c_0 + 2c_1 + 2c_2 + 2c_3 + \cdots \qquad (A.33)$$

In this case, two limits are worth noting. If the diagonal elements are much larger than any others, the eigenvalues are all essentially equal, of value $c_0$; if all the elements are nearly equal, there is one dominant eigenvalue of value approximately $mc_0$, and the remaining $(m-1)$ eigenvalues cluster around zero, often in an essentially singular manner.

To use these results on our competition matrices, we may for large $m$ pretend the resource spectrum is cyclic (so that the species labeled 0 adjoins that labeled $m-1$), whereupon matrices such as (6.15) are slightly modified to become members of the above class. In this way equation (A.32) leads to the specific result (6.17) for the matrix (6.15), and to the general result (6.22) for the general matrix (6.5) (May, 1972g). For large matrices, $m \gg 1$, the artifice is justified by the fact that "end effects" at the extremes of the resource spectrum are relatively unim-

portant. That this trick of imposing artificial cyclic boundary conditions does not affect the eigenvalues for $m \gg 1$ is
a point made clear in the literature on the physicists' Ising
model, where the trick is applied to linear chains or plane
lattices in exactly the same sense as here in Chapter 6. For a
more thorough discussion of the asymptotic validity of the
procedure, see for example Newell and Montroll (1953)
or Green and Hurst (1964). Further to substantiate this
point, consider the $m \times m$ matrix (A.29) with $c_0 = 1$, $c_1 =$
$c_{m-1} = \alpha$, and all other $c_i = 0$. This is just the matrix (A.26),
with the addition of an extra element $\alpha$ in both the upper
right-hand corner, and the lower left-hand corner (i.e. with
cyclic boundary conditions imposed). From (A.31) the
eigenvalues of this matrix are

$$\lambda_k = 1 + 2\alpha \cos (2\pi k/m),$$

with $k = 0, 1, \ldots, m - 1$, or equivalently (writing $j = 2k - m$)

$$\lambda_j = 1 - 2\alpha \cos \left(\frac{\pi j}{m}\right), \tag{A.34}$$

where $j$ runs over the even integers up to $2m$ if $m$ is even,
and over the odd integers to $2m - 1$ if $m$ is odd. Comparing
with the eigenvalues (A.28) of the matrix (A.26), we see the
eigenvalue distributions of the two matrices are significantly different for small $m$, but for large $m$ the expressions
(A.28) and (A.34) are effectively identical, and the easily
obtained result (A.34) serves as a good approximation for
the eigenvalues of the matrix (A.26)

# APPENDIX III

This appendix proves the assertions made in Chapter 2,
and illustrated by Figure 2.3, as to the relation between the

stability criteria for homologous pairs of differential and difference equations. The analysis aims to be of wider usefulness in indicating how the discussion of differential equation models throughout the book may in principle be taken over and applied to difference equation models in biological systems with discrete growth.

For the $m$-species community described by the general system of difference equations (2.25), the equilibrium populations are given by equation (2.9). To study the stability of this equilibrium point with respect to small disturbances, we as before write the perturbed populations as $N_i(t) = N_i^* + x_i(t)$ (equation (2.10)), and then obtain linearized equations for the initially small perturbations $x_i(t)$ by Taylor-expanding equations (2.25) around the equilibrium point:

$$x_i(t + \tau) - x_i(t) = \tau \sum_{j=1}^{m} a_{ij} x_j(t). \tag{A.35}$$

Here the coefficients $a_{ij}$ are defined by precisely the same recipe (2.13) as for the corresponding differential equations (2.8). Equation (A.35) may alternatively be expressed in matrix form (the analogue of equation (2.12)) as

$$\mathbf{x}(t + \tau) = B\mathbf{x}(t). \tag{A.36}$$

Clearly the elements $b_{ij}$ of the $m \times m$ matrix $B$ are

$$b_{ij} = \tau a_{ij} + \delta_{ij};$$

that is

$$B = \tau A + I. \tag{A.37}$$

In analogy with equation (2.14), for the set of linear difference equations (A.35) the solutions may be written

$$x_i(t) = \sum_{j=1}^{m} C_{ij} (\mu_j)^{t/\tau}. \tag{A.38}$$

The constants $C_{ij}$ are determined by the initial conditions, and the time dependence is contained solely in the $m$ factors $\mu^{t/\tau}$. The quantities $\mu_j$ are determined by substituting (A.38) into (A.36), noting that in effect $\mathbf{x}(t + \tau) = \mu\mathbf{x}(t)$:

$$\mu\mathbf{x}(t) = B\mathbf{x}(t).$$

Thence $\mu$ follows from the determinantal equation

$$\det |B - \mu I| = 0. \tag{A.39}$$

From equation (A.38) it is plain that a necessary and sufficient condition for initially small perturbations to die away in time is that all the quantities $\mu$ (which may be complex numbers) have magnitudes less then unity,

$$|\mu| < 1. \tag{A.40}$$

This stability condition on the eigenvalues of the matrix $B$ is, for the system of difference equations (2.25), precisely analogous to the condition (2.19), Real $(\lambda) < 0$, on the eigenvalues $\lambda$ of the matrix $A$ for the system of differential equations (2.8).

There is moreover a simple relationship between the eigenvalues $\mu$ of $B$, and the eigenvalues $\lambda$ of $A$. Using the identity (A.37), equation (A.39) becomes

$$\det |\tau A - (\mu - 1)I| = 0.$$

Comparison with equation (2.17) shows that

$$\mu = 1 + \tau\lambda. \tag{A.41}$$

The stability criterion (A.40) for the system with discrete generations thus reads

$$|\lambda + (1/\tau)| < (1/\tau). \tag{A.42}$$

This stability condition is as illustrated in Figure 2.3. As discussed in Chapter 2, it is to be contrasted with the less stringent criterion illustrated by Figure 2.1 for the corresponding system of differential equations.

# APPENDIX IV

In this appendix we take up some technical points associated with stochastic differential equations, and the definition of white noise. The coverage is very sketchy and is intended only to guide the interested reader to the relevant mathematical literature.

Definitions of continuous time white noise are to be found in Astrom (1970, pp. 30–33), Saaty (1967, Ch. 8), or Bartlett (1966, p. 243). Mathematically this noise spectrum is best regarded as the derivative of Brownian motion $(z(t))$, $\gamma dt = dz$, so that

$$\text{mean,} \quad \langle \gamma(t) \rangle = 0$$

$$\text{correlation,} \quad \langle \gamma(t)\gamma(t') \rangle = \sigma^2 \delta(t - t') \qquad (A.43)$$

with $\delta$ the usual Dirac delta-function. As discussed in Chapter 5, the correlation time for the fluctuations in any real system is finite; this point is covered in Astrom (1970, pp. 32–33), and with the addition of experimental tests of the validity of the mathematical theory by Morton and Corrsin (1969). In computer simulations, the continuous-time differential equations are replaced by a finite-difference approximation: $\gamma(t)$ is then to be interpreted as a gaussian noise spectrum, having no correlation from one time step to the next, and a variance $\sigma^2/(\Delta t)$, where $\Delta t$ is the finite-difference time step (e.g., Morton and Corrsin, equation (4)). Such an approximation will be warranted if the time step $\Delta t$ is small enough, which can be checked by verifying that one's results are unaltered when $\Delta t$ is, for example, halved.

The passage from a stochastic differential equation, incorporating white noise random fluctuations, to an equation for the concomitant probability distribution function has been treated by many people, with a variety of levels of sophistication. One simple and lucid account is by Kimura and Ohta (1971, Appendix I). More detailed discussions of the Fokker-Planck equations (5.2) and (5.7), and of their associated "friction" and "diffusion" coefficients, are in Astrom (1970, Chs. 3-7 and 3-8, complete with an excellent historical bibliography), Saaty (1967, Ch. 8), or Bartlett (1966, Ch. 3.5).

An important caution must be kept in mind. One must be circumspect about applying the rules of ordinary calculus to stochastic differential equations, as some recondite points are involved. For example, there are subtle differences in the formulae for the Fokker-Planck "friction" term depending on whether one first lets the correlation time for the fluctuations be zero and then takes the continuous-time limit, or whether one first takes continuous time and then lets the noise correlation time vanish. The consequent two different stochastic calculi are associated with the names of Ito and Stratonovich, respectively. The work in this book follows the usage of Ito, which is clearly appropriate to the way numerical computations are usually carried out (as for example in Figure 5.5), or to the case when continuous-time is used as an approximation to discrete growth steps. A good general account of the respective ground rules for these two stochastic calculi is to be found in the special review by Mortensen (1969); a more brief discussion is in Saaty (1967, Ch. 8.4).

In the approximate results discussed in Chapter 5 and Appendix V below, and forming the basis for most of Chapter 6, these complications fortunately do not arise. In this approximation, based on the assumption of relatively small fluctuations, both Ito's and Stratonovich's approaches lead to the same result. However, neglect of the subtleties of stochastic calculus has on occasion lead to errors in the ecological literature.

## APPENDIX V

This appendix indicates some methods of solving the time-independent Fokker-Planck equation, to obtain equilibrium probability distributions, $f^*$. We first treat one-dimensional (single species) equations exactly, and then in the multidimensional case we discuss a widely useful approximation.

### (i) One-dimensional Equilibrium Distribution

For a single-species population, the equilibrium probability distribution function $f^*(n)$, if it exists, is given by the time-independent limit of equation (5.2):

$$0 = -\frac{\partial}{\partial n} (M(n)f^*(n)) + \frac{1}{2} \frac{\partial^2}{\partial n^2} (V(n)f^*(n)). \quad \text{(A.44)}$$

Here the coefficients $M(n)$ and $V(n)$ are obtained from the stochastic differential equation (5.1) by the prescriptions (5.3) and (5.4). This equation can be integrated to give (see, e.g., Wright, 1938 or Bartlett, 1966, Ch. 4.3)

$$f^*(n) = \frac{C}{V(n)} \exp \left[ 2 \int^n \frac{M(n')}{V(n')} \, dn' \right]. \quad \text{(A.45)}$$

$C$ is just the appropriate normalization constant. If the function (A.45) is not well behaved, it may in general be inferred that there is no equilibrium distribution.

The distribution (5.12) of Chapter 5 follows from the general result. In this instance we have from equation (5.10) that $M(n) = n(k_o - n)$ and $V(n) = \sigma^2 n^2$, and (A.45) leads directly to (5.12). This probability distribution is integrable only if the limiting behavior $n^\alpha$ in the neighbor-hood of the origin has $\alpha > -1$; that is, the equilibrium probability distribution only exists if the restriction (5.13) is obeyed.

Notice that in Fokker-Planck equations such as (A.44) the population variable $n$ is taken to be continuous. Then the probability to find $n$ within some infinitesimal interval $\epsilon$ of the origin can tend to zero as $\epsilon \to 0$, and a well-defined stationary probability distribution can exist. In nature, animals come quantized in integer values, and the probability to have exactly zero animals (extinction) is estimated by the integral over $f^*(n)$ from $n = 0$ to $n = 1$. This will be some finite number, however small. The consequent "probability leakage" or "absorbing barrier" at $n = 0$ means that in the long run there may be no stationary distribution. But if this extinction probability is very small, as it usually will be if the average population is very large, such complications will be irrelevant on any biologically reasonable time scale, and our discussion of the equilibrium probability distribution is sensible. (By way of illustration, in our example of the stochastic logistic equation (5.10), if $k_o = 1,000$ and $\sigma^2/k_o = 0.2$, the probability to find less than one animal is around $10^{-24}$.)

### (ii) Approximate Multidimensional Equilibrium Distribution

Much of the discussion in Chapter 5 drew upon an approximate solution for the equilibrium probability distribution which applies when the fluctuations are not too severe, that is when the probability cloud is relatively compact.

Allowing for random fluctuations in the parameters of the model, the stochastic generalization of the system (2.8) may be written

$$\frac{dN_i(t)}{dt} = F_i(N_1(t), \ldots, N_m(t); \{\kappa_j(t)\}) \qquad (A.46)$$

where each of the set of interaction parameters $\{\kappa(t)\}$ has the form

$$\kappa_j(t) = \kappa_j^0 + \gamma_j(t). \qquad (A.47)$$

Here $\kappa^0$ is the constant average value of the parameter, and $\gamma(t)$ is white noise, with the covariance between the $i$th and $j$th environmental fluctuation being measured by some $\sigma_{ij}^2$. Mean populations $N_i^*$ may be defined as the solutions of the time-independent equations (2.9), using the mean values $\{\kappa_j^0\}$ for the interaction parameters. The actual fluctuating populations may be expanded about these constant mean values as in Chapter 2: $N_i(t) = N_i^* + x_i(t)$. The quantities $x_i$ measure the population fluctuations; alternatively the *relative* fluctuations may be expressed by $\chi_i(t) \equiv x_i(t)/N_i^*$. For relatively small environmental fluctuations, $\sigma_{ij}$ relatively small, the equations (A.46) may be Taylor-expanded about this mean point, keeping only terms of first order in $\chi_i$ and in the parameter fluctuations $\gamma_i$. This leads to the linear stochastic differential equations

$$\frac{d\chi_i(t)}{dt} = \sum_{j=1}^{m} ((N_i^*)^{-1}a_{ij}N_j^*)\chi_j(t) + \sum_k \mu_{ik}\gamma_k(t). \qquad (A.48)$$

Here the $a_{ij}$ are calculated exactly as for the deterministic problem, namely from equation (2.13), using the average populations and average interaction parameters. In this sense, this (average) community matrix is just the (deterministic) community matrix of Chapter 2. The overall coefficients $\bar{a}_{ij} = (N_i^*)^{-1}a_{ij}N_j^*$ form the matrix $\bar{A}$ whose definition and properties are discussed in the main text (equation (5.31) et seq.). In a similar way, the constant coefficients $\mu_{ij}$ involve the average populations and parameters.

The approximate differential equation (A.48), with its

white noise stochasticity, leads via the recipes (5.7) with (5.8) and (5.9) to a Fokker-Planck equation for the multivariate probability distribution function. The approximate equilibrium distribution $\hat{f}*(\nu_1, \nu_2, \ldots, \nu_m)$, if it exists, gives the probability to observe the relative fluctuations having the values $\chi_i(t) = \nu_i$ (in accordance with the definitions (5.16), (5.24), and (5.28)). It is determined by the equation

$$0 = -\sum_{i,j} \frac{\partial}{\partial \nu_i} (\bar{a}_{ij}\nu_j \hat{f}*) + \frac{1}{2} \sum_{i,j} D_{ij} \frac{\partial^2(\hat{f}*)}{\partial \nu_i \partial \nu_j}. \quad (A.49)$$

Here $D_{ij}$ formally represents the overall covariance between the white noise fluctuations in the stochastic differential equation for the $i$th population and that for the $j$th.

As may be seen by direct substitution, the solution of equation (A.49) is the multivariate gaussian distribution of equation (5.29), namely

$$\hat{f}*(\nu_1, \nu_2, \ldots, \nu_m) = C \exp\left[\sum_{ij} \nu_i B_{ij}^{-1} \nu_j\right], \quad (5.29)$$

where the elements of the symmetric covariance matrix, $B_{ij}$, must obey the equations

$$\sum_k (B_{ik}^{-1} \bar{a}_{kj} + \bar{a}_{ik}^T B_{kj}^{-1}) = 2 \sum_{k,\ell} B_{ik}^{-1} D_{k\ell} B_{\ell j}^{-1}.$$

This immediately gives the Lyapunov matrix equation (5.30) relating $B$, $\bar{A}$, and $D$. The properties of the approximate distribution function (5.29) are as discussed in Chapter 5.

Barnett and Storey (1970, Ch. 6) and Astrom (1970, Ch. 3-6) contain somewhat different and considerably more elaborate expositions along the above lines.

So far the discussion has been couched in general terms, as pertains to the general multispecies models in Chapter 5 (pp. 130–133). It may help to set out a specific application. Consider the stochastic equations (A.46) for two competing species. Here equations (5.19) with (5.23) are a particular realization of the general system (A.46), with

$$F_1 = N_1(k_o + \gamma_1(t) - N_1 - \alpha N_2)$$
$$F_2 = N_2(k_o + \gamma_2(t) - N_2 - \alpha N_1).$$

According to the above prescription, the $2 \times 2$ matrix $\bar{A}$ is

$$\bar{A} = \begin{pmatrix} -N_1^* & -\alpha N_2^* \\ -\alpha N_1^* & -N_2^* \end{pmatrix}.$$

Or, remembering the notation of equation (5.21), $N^* = k_o/(1 + \alpha)$,

$$\bar{A} = -N^* \begin{pmatrix} 1 & \alpha \\ \alpha & 1 \end{pmatrix}.$$

The white noise covariance matrix $D$ in this example is particularly simple, as we assumed no correlation between the noise spectra $\gamma_1(t)$ and $\gamma_2(t)$:

$$D = \begin{pmatrix} \sigma^2 & 0 \\ 0 & \sigma^2 \end{pmatrix}.$$

Since $\bar{A}$ is symmetric, the Lyapunov matrix equation (5.30) reduces to

$$B^{-1} = D^{-1}\bar{A}.$$

The matrix $D$ is trivially inserted here, leading to the final result

$$B^{-1} = -\frac{N^*}{\sigma^2} \begin{pmatrix} 1 & \alpha \\ \alpha & 1 \end{pmatrix}.$$

Whence the specific result (5.25) follows from the general expression (5.29). Even more easily we could derive the approximate result (5.17) for the single-species stochastic system (5.10).

To reiterate, this approximation is valid if the probability cloud is not too diffuse. As such, it contributes useful results to Chapters 5 and 6, particularly the rough stability criterion illustrated by Figure 5.2. Once the fluctuations in the population numbers become very severe, the approximation is inappropriate.

# Bibliography

Allee, W. C. 1939. *The social life of animals*. London: Heinemann.

Aruego, J. 1970. *Symbiosis: a book of unusual friendships*. New York: Charles Scribner.

Ashby, W. R. 1952. *Design for a brain*. London: Chapman and Hall.

Astrom, K. J. 1970. *Introduction to stochastic control theory*. New York: Academic Press.

Baltensweiler, W. 1971. The relevance of changes in the composition of larch bud moth populations for the dynamics of its numbers. In P. J. den Boer and G. R. Gradwell (eds.), *Dynamics of populations*. pp. 208–219. Wageningen: Centre for agricultural publishing and documentation.

Barnett, S., and Storey, C. 1970. *Matrix methods in stability theory*. London: Nelson.

Bartlett, M. S. 1960. *Stochastic population models*. London: Methuen & Co.

Bartlett, M. S. 1966. *An introduction to stochastic processes*. 2nd ed. Cambridge: Cambridge University Press.

Beaver, D. 1972. Private communication. (Publication in preparation.)

Becker, N. G. 1973. Interactions between species: some comparisons between deterministic and stochastic models. *Rocky Mountains J. Math., 3*, in press.

Berlin, T. H., and Kac, M. 1952. The spherical model of a ferromagnet. *Phys. Rev., 86*, 821–835.

Beverton, R. J. H., and Holt, S. J. 1956. The theory of fishing. In M. Graham (ed.), *Sea fisheries, their investigation in the United Kingdom,* pp. 372–441. London: Edward Arnold Ltd.

Boorman, S. A., and Levitt, P. R. 1972. Group selection on the boundary of a stable population. *Proc. Nat. Acad. Sci., 69,* 2711–2713.

Bulgakova, T. I. 1968a. Concerning the stability of the simplest model of biogeocenosis. *Problemy Kibernetiki, 20,* 271–276. (See Translations in the field of Soviet cybernetics: U.S. Dept. of Commerce. JPRS 48396, pp. 59–65, 1969).

Bulgakova, T. I. 1968b. Concerning models of the competition of species. *Ibid.,* 263–270. (See Translations, as above, pp. 50–58).

Burns, J. 1970. Comment on Kauffman 1970b. In C. H. Waddington (ed.), *Towards a theoretical biology: 3. Drafts,* pp. 44–45. Edinburgh: Edinburgh University Press.

Burton, T. A. 1969. On the construction of Lyapunov functions. *SIAM J. Appl. Math., 17,* 1078–1085.

Canale, E. P. 1970. An analysis of models describing predator-prey interaction. *Biotech. and Bioeng., 12,* 353–378.

Clark, L. R., Geier, P. W., Hughes, R. D., and Morris, R. F. 1967. *The ecology of insect populations in theory and practice.* London: Methuen and Co.

Cody, M. 1968. On the methods of resource division in grassland bird communities. *Amer. Natur., 102,* 107–148.

Cole, L. C. 1954. Some features of random cycles. *J. Wildlife Mgmt., 18,* 107–109.

Connell, J. H., and Orians, E. 1964. The ecological regulation of species diversity. *Amer. Nat., 98,* 399–414.

Connell, J. H. 1971. On the role of natural enemies in preventing competitive exclusion in some marine animals and in rain forest trees. In *Dynamics of populations. op. cit.,* pp. 298–312.

212

Conway, G. R. 1970. Computer simulation as an aid to developing strategies for *Anopheline* control. *Misc. Publ. Entomol. Soc. Amer., 7,* 181–193.

Conway, G. R. 1971. Better methods of pest control. In W. W. Murdoch (ed.), *Environment.* Stamford: Sinauer Assoc., pp. 302–325.

Conway, G. R. 1972. Experience in insect pest modelling: a review of models, uses and future directions. In *Proceedings XIV International Entomological Congress.* Canberra: Proc. Ecol. Soc. Aust. (in press).

Conway, G. R., and Murdie, G. 1972. Population models as a basis for pest control. In J. N. R. Jeffers (ed.), *Mathematical models in ecology.* pp. 195–213. Oxford: Blackwell Scientific Publications.

Cramer, N. F., and May, R. M. 1971. Interspecific competition, predation and species diversity: a comment. *J. Theor. Biol., 34,* 289–293.

Curtis, J. T. 1956. The modification of mid-latitude grasslands and forests by man. In W. L. Thomas (ed.), *Man's role in changing the face of the earth,* pp. 721–736. Chicago: University of Chicago Press.

Curtis, J. T. 1959. *The vegetation of Wisconsin: an ordination of plant communities.* Madison, Wisconsin: Wisconsin University Press.

Dale, M. B. 1970. Systems analysis and ecology. *Ecology, 51,* 2–16.

Darlington, P. J. 1957. *Zoogeography.* New York: John Wiley and Sons.

Darlington, P. J. 1965. *Biogeography of the southern end of the world.* New York: McGraw-Hill.

Deakin, M. A. B. 1971. Restrictions on the applicability of Volterra's ecological equations. *Bull. Math. Biophys., 33,* 571–578.

de Bach, P. (ed.) 1964. *Biological control of insect pests and weeds,* pp. 124–128; elsewhere. New York: Reinhold.

213

den Boer, P. J. 1968. Spreading of risk and stabilisation of animal numbers. *Acta Biotheor.* (Leiden), *18*, 165–194.

Demetrius, L. 1969. On community stability. *Math. biosci.,* *5*, 321–325.

Diamond, J. M. 1972. *The avifauna of the eastern highlands of New Guinea.* Cambridge, Mass.: Publ. Nuttall Ornithol. Club.

Dixon, K. R., and Cornwell, G. W. 1970. A mathematical model for predator and prey populations. *Res. Pop. Ecol., 12,* 127–136.

Dumas, P. C. 1956. The ecological relations of sympatry in *Plethodon dunni* and *P. vehiculum. Ecology, 37,* 484–495.

Dunkel, G. M. 1968a. Single species model for population growth depending on past history. In *Seminar on differential equations and dynamic systems,* lecture notes in mathematics, Vol. 60. Heidelberg: Springer-Verlag, pp. 92–99.

Dunkel, G. M. 1968b. Some mathematical models for population growth with lags. University of Maryland, Inst. Fluid Dynamics and App. Math., Tech. Note BN-548.

Ehrlich, P. R., and Birch, L. C. 1967. The "Balance of nature" and "Population control." *Amer. Natur., 101,* 97–107.

Elton, C. S. 1958. *The ecology of invasions by animals and plants.* London: Methuen and Co.

Elton, C. S. 1966. *The pattern of animal communities.* London: Methuen & Co.

Emerson, A. E. 1949. In W. C. Allee, A. E. Emerson, O. Park, T. Park, and K. P. Schmidt (eds.), *Principles of animal ecology,* pp. 640–695. Philadelphia: Saunders.

Frisch, R., and Holme, H. 1935. The characteristic solutions of a mixed difference and differential equation occurring in economic dynamics. *Econometrica, 3,* 225–239.

Gardner, M. R., and Ashby, W. R. 1970. Connectance of large dynamical (cybernetic) systems: critical values for stability. *Nature, 228,* 784.

Garfinkel, D. A. 1967. A simulation study of the effects on simple ecological systems of making rate of increase of population density dependent. *J. Theor. Biol., 14,* 46–58.

Gause, G. F. 1934. *The struggle for existence.* Baltimore: Williams and Wilkins.

Gilpin, M. E. 1972. Enriched predator-prey systems: theoretical stability. *Science, 177,* 902–904.

Goel, N. S., Maitra, S. C., and Montroll, E. W. 1971. On the Volterra and other nonlinear models of interacting populations. *Rev. Mod. Phys., 43,* 231–276.

Gómez-Pompa, A., Vazquez-Yanes, C., and Guevara, S. 1972. The tropical rain forest: a nonrenewable resource. *Science, 177,* 762–765.

Gompertz, B. 1825. On the nature of the function expressive of the law of human mortality. *Phil. Trans., 115,* 513–585.

Goodwin, B. 1970. Biological Stability. In *Towards a Theoretical Biology: 3. Drafts, op. cit.,* pp. 1–17.

Grant, P. R. 1966. Ecological compatibility of bird species on islands. *Amer. Natur., 100,* 451–462.

Green, H. S., and Hurst, C. A. 1964. *Order-disorder phenomena.* London: Interscience Publishers.

Hairston, N. G., Smith, F. E., and Slobodkin, L. B., 1960. Community structure, population control, and competition. *Amer. Natur., 94,* 421–425.

Hairston, N. G., et al. 1968. The relationship between species diversity and stability: an experimental approach with protozoa and bacteria. *Ecology, 49,* 1091–1101.

Hall, D. J., Cooper, W. E., and Werner, E. E. 1970. An experimental approach to the production dynamics and

215

structure of freshwater animal communities. *Limnol. Oceanogr., 15,* 839–928.

Hassell, M. P., and Varley, G. C. 1969. New inductive population model for insect parasites and its bearing on biological control. *Nature, 223,* 1133–1137.

Hassell, M. P., and Rogers, D. J. 1972. Insect parasite responses in the development of population models. *J. Anim. Ecol., 41,* 661–676.

Hassell, M. P., and May, R. M. 1973. Stability in insect host-parasite models. (To be published.)

Heatwole, H., and Davis, D. M. 1965. Ecology of three sympatric species of parasitic insects of the genus *megarhyssa. Ecology, 46,* 140–150.

Hespenheide, H. A. 1971. Food preference and the extent of overlap in some insectivorous birds, with special reference to the *Tyrannidae. Ibis, 113,* 59–72.

Ho, T-Y. 1967. Relationship between imunoacid contents of mammalian bone collagen and body temperature, as a basis for estimation of body temperature of prehistoric mammals. *Comp. Biochem. Physiol., 22,* 113–119.

Holdgate, M. W., and Wace, N. M. 1971. The influence of man on the floras and faunas of southern islands. In T. R. Detwyler (ed.), *Man's impact on environment,* pp. 476–492. New York: McGraw-Hill.

Holling, C. S. 1959. The components of predation as revealed by a study of small-mammal predation of the European pine sawfly. *Canad. Ent., 91,* 293–320.

Holling, C. S. 1961. Principles of insect predation. *Ann. Rev. Entomol., 6,* 163–182.

Holling, C. S. 1965. The functional response of predators to prey density and its role in mimicry and population regulation. *Mem. Entomol. Soc. Can., 45,* 1–60.

Holling, C. S. 1966. The strategy of building models of complex ecological systems. In Watt, *op. cit.,* pp. 195–214.

216

Holling, C. S. 1968a. The tactics of a predator. In T. R. E. Southwood (ed.), *Insect abundance*, p. 47. Oxford: Blackwell Scientific Publications.

Holling, C. S. 1968b. Personal communication quoted in Watt, *op. cit.*, p. 44.

Horn, H. S., and MacArthur, R. H. 1972. On competition in a diverse and patchy environment. *Ecology, 53,* 749–752.

Hubble, S. P. 1973. Populations and simple food webs as energy filters: II, Two-species systems. *Amer. Natur., 107,* 122–151.

Huffaker, C. B. 1958. Experimental studies on predation: dispersion factors and predator-prey oscillations. *Hilgardia, 27,* 343–383.

Hutchinson, G. E. 1948. Circular causal systems in ecology. *Ann. N.Y. Acad. Sci., 50,* 221–246.

Hutchinson, G. E. 1954. Theoretical notes on oscillatory populations. *J. Wildlife Mgmt., 18,* 107–109.

Hutchinson, G. E. 1957. Concluding remarks. *Cold Spring Harb. Symp. Quant. Biol., 22,* 415–427.

Hutchinson, G. E. 1959. Homage to Santa Rosalia, or why are there so many kinds of animals? *Amer. Natur., 93,* 145–159.

Hutchinson, G. E. 1961. The paradox of the plankton. *Amer. Natur., 95,* 137–145.

Ivlev, V. S. 1961. *Experimental ecology of the feeding of fishes.* New Haven: Yale University Press.

Janzen, D. H. 1970. Herbivores and the number of tree species in tropical forests. *Amer. Natur., 104,* 501–528.

Jones, G. S. 1962a. The existence of periodic solutions of $f'(x) = -\alpha f(x-1)[1+f(x)]$. *J. Math. Anal. Appl., 5,* 435–450.

Jones, G. S. 1962b. On the nonlinear differential difference equation $f'(x) = -\alpha f(x-1)[1+f(x)]$. *J. Math. Anal. Appl., 4,* 440–469.

Kaplan, J. L., and Yorke, J. A. 1973. On the stability of a periodic solution of a differential-delay equation. *SIAM J. Math. Anal.*, in press.

Kauffman, S. 1970a. Behaviour of randomly constructed genetic nets: binary element nets. In *Towards a Theoretical biology: 3. Drafts, op. cit.*, pp. 18–37.

Kauffman, S., 1970b. Behaviour of randomly constructed genetic nets: continuous element nets. *Ibid.*, pp. 38–44.

Kerner, E. H. 1957. A statistical mechanics of interacting biological species. *Bull. Math. Biophys.*, *19*, 121–146.

Kerner, E. H. 1959. Further considerations on the statistical mechanics of biological associations. *Bull. Math. Biophys.*, *21*, 217–255.

Kerner, E. H. 1961. On the Volterra-Lotka principle. *Bull. Math. Biophys.*, *23*, 141–157.

Kerner, E. H. 1969. Gibbs ensemble and biological ensemble. In C. H. Waddington (ed.), *Towards a theoretical biology: 2. Sketches*, pp. 129–139. Edinburgh: Edinburgh University Press.

Kilmer, W. L. 1972. On some realistic constraints in prey-predator mathematics. (To be published.)

Kimura, M., and Ohta, T. 1971. *Theoretical aspects of population genetics.* Princeton: Princeton University Press.

Klopfer, P. H. 1962. *Behavioural aspects of ecology.* Englewood Cliffs, N.J.: Prentice-Hall.

Kolomogorov, A. N. 1936. Sulla Teoria di Volterra della Lotta per l'Esisttenza. *Giorn. Instituto Ital. Attuari, 7,* 74–80.

Kormondy, E. J. 1969. *Concepts of ecology.* Englewood Cliffs, N.J.: Prentice-Hall.

Kurten, B. 1969. Continental drift and evolution. *Sci. Amer., 220,* 54–65.

Lack, D. L. 1954. *The natural regulation of animal numbers.* Oxford: Clarendon Press.

Leigh, E. 1965. On the relation between the productivity,

biomass, diversity, and stability of a community. *Proc. Nat. Acad. Sci.*, *53*, 777–783.

Leigh, E. 1968. The ecological role of Volterra's equations. In *Some mathematical problems in biology.* Providence: The American Mathematical Society.

Leigh, E. 1971. *Adaptation and diversity.* San Francisco: Freeman, Cooper and Co., Ch. 10.

Leslie, P. H. 1948. Some further notes on the use of matrices in population mathematics. *Biometrica, 35,* 213–245.

Leslie, P. H., and Gower, J. C. 1960. The properties of a stochastic model for the predator-prey type of interaction between two species. *Biometrica, 47,* 219–234.

Levin, S. 1970. Community equilibria and stability, and an extension of the competitive exclusion principle. *Amer. Natur., 104,* 413–423.

Levins, R. 1966. The strategy of model building in population biology. *Amer. Sci., 54,* 421–431.

Levins, R. 1968a. *Evolution in changing environments.* Princeton: Princeton University Press.

Levins, R. 1968b. Ecological engineering: theory and practice. *Quart. Rev. Biol., 43,* 301–305.

Levins, R. 1969a. Some demographic and genetic consequences of environmental heterogeneity for biological control. *Bull. Entomol. Soc. Am., 15,* 237–240.

Levins, R. 1969b. The effect of random variations of different types on population growth. *Proc. Nat. Acad. Sci., 62,* 1061–1065.

Levins, R. 1970a. Extinction. In M. Gerstenhaber (ed.), *Some mathematical problems in biology* (Vol. II of Lectures on Mathematics in the Life Sciences), pp. 77–107. Providence: The American Mathematical Society.

Levins, R. 1970b. Complex systems. In *Towards a theoretical biology: 3. Drafts, op. cit.,* pp. 73–88.

Lewontin, R. C. 1969. The meaning of stability. In *Diversity and stability in ecological systems,* Brookhaven Symposium in Biology No. 22. Springfield Va.: National Bureau of Standards, U.S. Department of Commerce, pp. 13–24.

Lewontin, R. C., and Cohen, D. 1969. On population growth in a randomly varying environment. *Proc. Nat. Acad. Sci., 62,* 1056–1060.

Lotka, A. J. 1925. *Elements of physical biology.* Baltimore: Williams and Wilkins. (Reissued as *Elements of mathematical biology* by Dover, 1956.)

Lotka, A. J. 1932. Growth of mixed populations. *J. Wash. Acad. Sci., 22,* 461–469.

MacArthur, R. H. 1955. Fluctuations of animal populations, and a measure of community stability. *Ecology, 36,* 533–536.

MacArthur, R. H., and Levins, R. 1964. Competition, habitat selection, and character displacement in a patchy environment. *Proc. Nat. Acad. Sci., 51,* 1207–1210.

MacArthur, R. H., and Connell, J. H. 1966. *The biology of populations.* Ch. 3.6. New York: John Wiley and Sons.

MacArthur, R. H., and Pianka, E. S. 1966. On optimal use of a patchy environment. *Amer. Natur., 100,* 603–609.

MacArthur, R. H., and Levins, R. 1967. The limiting similarity, convergence, and divergence of coexisting species. *Amer. Natur., 101,* 377–385.

MacArthur, R. H., and Wilson, E. O. 1967. *The theory of island biogeography.* Princeton: Princeton University Press.

MacArthur, R. H. 1968. The theory of the niche. In *Population biology and evolution,* pp. 159–176. Syracuse: Syracuse University Press.

MacArthur, R. H. 1969. Species packing, or what competition minimizes. *Proc. Nat. Acad. Sci., 64,* 1369–1375.

MacArthur, R. H. 1970. Species packing and competitive equilibrium for many species. *Theor. Pop. Biol., 1,* 1–11.

MacArthur, R. H. 1971a. Graphical analysis of ecological systems. Division of Biology Report, Princeton University.

MacArthur, R. H. 1971b. Patterns of terrestrial bird communities. In *Avian biology*, Vol. I, pp. 189–221. New York: Academic Press.

MacArthur, R. H. 1972. *Geographical ecology*. New York: Harper & Row.

Macfadyen, A. 1963. *Animal ecology: aims and methods*. 2nd ed. London: Pitman and Sons.

McMurtrie, R. 1972. Numerical studies of the eigenvalues of random real matrices. (Publication in preparation.)

McNeil, D. R., and Schach, S. 1971. Central limit analogues for Markov population processes. Princeton University, Dept. Math. Stats. Technical Report No. 4.

Maelzer, D. A. 1970. The regression of $\log N(n + 1)$ on $\log N(n)$ as a test of density dependence. *Ecology, 51,* 810–822.

Margalef, R. 1968. *Perspectives in ecological theory*. Chicago: University of Chicago Press.

Martin, P. S. 1966. Africa and Pleistocene overkill. *Nature, 212,* 339–342.

May, R. M. 1971. Stability in multi-species community models. *Math. Biosci., 12,* 59–79.

May, R. M. 1972a. On relationships among various types of population models. *Amer. Natur., 107,* 46–57.

May, R. M. 1972b. Will a large complex system be stable? *Nature, 238,* 413–414.

May, R. M. 1972c. Mass and energy flow in closed ecosystems: a comment. *J. Theor. Biol.,* in press.

May, R. M. 1972d. Limit cycles in predator-prey communities. *Science, 177,* 900–902.

May, R. M. 1972e. Time-delay versus stability in population models with two and three trophic levels. *Ecology,* in press.

May, R. M. 1972f. Stability in randomly fluctuating versus deterministic environments. *Amer. Natur.*, in press.

May, R. M. 1972g. Theory of niche overlap in a fluctuating environment. *Theor. Pop. Biol.*, in press.

May, R. M., and MacArthur, R. H. 1972. Niche overlap as a function of environmental variability. *Proc. Nat. Acad. Sci.*, *69*, 1109–1113.

May, R. M. 1973. Ecological systems in randomly fluctuating environments. In R. Rosen and F. Snell (eds.), *Progress in theoretical biology*. New York: Academic Press. (To be published.)

Maybee, J. S., and Quirk, J. P. 1969. Qualitative problems in matrix theory. *SIAM Rev.*, *11*, 30–51.

Maynard Smith, J. 1968. *Mathematical ideas in biology*. Cambridge: Cambridge University Press.

Maynard Smith, J. 1971. Private communication. (Publication in preparation.)

Mech, L. D. 1966. *The wolves of Isle Royle*. U.S. Nat. Park Serv.: Fauna Nat. Parks U.S., Fauna Series No. 7.

Miller, R. S. 1967. Pattern and process in competition. In J. B. Cragg (ed.), *Advances in ecological research*. Vol. 4, pp. 1–74. New York: Academic Press.

Minorsky, N. 1962. *Non-linear oscillations*. Princeton: Van Nostrand.

Moore, R. C. (ed.) 1953–1967. *Treatise on invertebrate paleontology*. (13 Vols.) Kansas: Univ. Kansas Press. Part L4.

Morowitz, H. J. 1968. *Energy flow in biology*. New York: Academic Press.

Mortensen, R. E. 1969. Mathematical problems of modeling stochastic non-linear dynamic systems. *J. Stat. Phys.*, *1*, 271–296.

Morton, J. B., and Corrsin, S. 1969. Experimental confirmation of the applicability of the Fokker-Planck equation to a non-linear oscillator. *J. Math. Phys.*, *10*,

361–368.

Murdoch, W. W. 1966. "Community structure, population control, and competition"—a critique. *Amer. Natur.*, *100*, 219–226.

Murdoch, W. W. 1969. Switching in general predators: experiments on predator specificity and stability of prey populations. *Ecol. Monogr.*, *39*, 335–354.

Newell, G. F., and Montroll, E. W. 1953. On the theory of the Ising model of ferromagnetism. *Rev. Mod. Phys.*, *25*, 353–389.

Nicholson, A. J. 1954. An outline of the dynamics of animal populations. *Aust. J. Zoo.*, *2*, 9–65.

Nicholson, A. J. 1957. The self-adjustment of populations to change. *Cold Spring Harb. Symp. Quant. Biol.*, *22*, 153–173.

Odum, E. P. 1953. *Fundamentals of ecology.* Philadelphia-London: W. B. Saunders.

Paine, R. T. 1966. Food web complexity and species diversity. *Amer. Natur.*, *100*, 65–75.

Parrish, J. D., and Saila, S. B. 1970. Interspecific competition, predation, and species diversity. *J. Theor. Biol.*, *27*, 207–220.

Patten, B. C. (ed.). 1971. *Systems analysis and simulation in ecology.* Vol. I. New York: Academic Press.

Pianka, E. R. 1966. Latitudinal gradients in species diversity: a review of concepts. *Amer. Natur.*, *100*, 33–46.

Pielou, E. C. 1969. *An introduction to mathematical ecology.* New York: Wiley-Interscience.

Pimentel, D. 1961. Species diversity and insect population outbreaks. *Ann. Entomol. Soc. Am.*, *54*, 76–86.

Quirk, J. P., and Ruppert, R. 1965. Qualitative economics and the stability of equilibrium. *Rev. Econ. Studies*, *32*, 311–326.

Raup, D. M. 1972. Taxonomic diversity during the Phanerozoic. *Science*, *177*, 1065–1071.

Reddingius, J., and den Boer, P. J. 1970. Simulation experiments illustrating stabilization of animal numbers by spreading of risk. *Oecologia, 5,* 240–284.

Rescigno, A., and Richardson, I. W. 1965. On the competitive exclusion principle. *Bull. Math. Biophys., 27,* 85–89.

Rescigno, A., and Richardson, I. W. 1967. The struggle for life: I, two species. *Bull. Math. Biophys., 29,* 377–388.

Rescigno, A. 1968. The struggle for life: II, three competitors. *Bull. Math. Biophys., 30,* 291–298.

Roff, D. 1972. Spatial heterogeneity and the persistence of populations. (To be published.)

Romer, A. S. 1966. *Vertebrate paleontology.* 3rd ed. Chicago: Chicago Univ. Press.

Rosen, R. 1970. *Dynamical system theory in biology.* Vol. I. New York: John Wiley and Sons.

Rosenzweig, M., and MacArthur, R. H. 1963. Graphical representation and stability conditions of predator–prey interaction. *Amer. Natur., 97,* 209–223.

Rosenzweig, M. 1971. Paradox of enrichment: destabilisation of exploitation ecosystems in ecological time. *Science, 171,* 385–387.

Roughgarden, J. 1972. The evolution of niche width. *Amer. Natur., 106,* 683–718.

Roughgarden, J. 1973. On invading a guild of competing species. *Amer. Natur.,* in press.

Saaty, T. L. 1967. *Modern non-linear equations.* New York: McGraw-Hill.

St. Amant, J. L. S. 1970. The detection of regulation in animal populations. *Ecology, 51,* 823–828.

St. Amant, J. L. S. 1971. Notes on the mathematics of predator–prey interactions. M.A. Thesis, University of California at Santa Barbara.

Schoener, T., and Gorman, G. 1968. Some niche differences in three lesser antillean lizards of the genus *Anolis. Ecology, 49,* 819–830.

224

Schoener, T. W. 1969. Optimal size and specialization in constant and fluctuating environments: an energy-time approach. In *Diversity and stability in ecological systems, op. cit.*, pp. 103–114.

Schoener, T. W. 1972. Population growth regulated by intraspecific competition for energy or time, in press.

Schoener, T. W. 1973. Competition and the form of habitat shift. (To be published.)

Scudo, F. M. 1971. Vito Volterra and theoretical ecology. *Theor. Pop. Biol., 2,* 1–23.

Simpson, G. G. 1953. *Evolution and geography: an essay on historical biogeography with special reference to mammals.* Eugene, Oregon: Oregon University Press.

Simpson, G. G. 1964. Species density of North American recent mammals. *System. Zoo., 13,* 57–73.

Simpson, G. G. 1965. *The geography of evolution.* Philadelphia: Chilton.

Simpson, G. G. 1969. The first three billion years of community evolution. In *Diversity and stability in ecological systems, op. cit.*, pp. 162–177.

Skellam, J. G. 1951. Random dispersal in theoretical populations. *Biometrica, 38,* 196–218.

Slobodkin, L. B. 1961. *Growth and regulation of animal populations.* New York: Holt, Rinehart and Winston.

Slobodkin, L. B., Smith, F. E., and Hairston, N. G. 1967. Regulation in terrestrial ecosystems, and the implied balance of nature. *Amer. Natur., 101,* 109–124.

Slobodkin, L. B., and Sanders, H. L. 1969. On the contribution of environmental predictability to species diversity. In *Diversity and stability in ecological systems, op. cit.*, pp. 82–95.

Slone, N. J. H. 1967. *Lengths of cycle time in random neural networks.* Ithaca: Cornell University Press.

Smale, S. 1966. Structurally stable systems are not dense. *Am. J. Math., 87,* 491–496.

Smith, F. E. 1963. Population dynamics in *Daphnia Magna* and a new model for population growth. *Ecology, 44,* 651–663.

Smith, F. E. 1969. Effects of enrichment in mathematical models. In *Eutrophication: causes, consequences, correctives.* Washington D.C.; National Academy of Sciences, pp. 631–645.

Smith, F. E. 1972. Spatial heterogeneity, stability, and diversity in ecosystems. *Trans. Conn. Acad. Arts and Sci., 44,* 307–335.

Southwood, T. R. E. 1961. The number of species of insect associated with various trees. *J. Anim. Ecol., 30,* 1–8.

Southwood, T. R. E., and Way, M. J. 1970. Ecological background to pest management. In R. L. Rabb and F. E. Guthrie (eds.), *Concepts of pest management,* pp. 6–28. Raleigh: North Carolina State University Press.

Storer, R. W. 1966. Sexual dimorphism and food habits in three North American accipiters. *Auk., 83,* 423–436.

Sykes, Z. M. 1969. Some stochastic versions of the matrix model for population dynamics. *J. Am. Stat. Assoc., 64,* 111–130.

Takahashi, F. 1964. Reproduction curve with two equilibrium points: A consideration on the fluctuation of insect population. *Res. Pop. Ecol., 6,* 28–36.

Tanner, J. T. 1966. Effects of population density on growth rates of animal populations. *Ecology, 47,* 733–745.

Tanner, J. T. 1972. The stability and the intrinsic growth rates of prey and predator populations. (To be published.)

Terborgh, J. 1972. In preparation. Results quoted in MacArthur, 1972, *op. cit.:* see Figure 2-6.

Thom, R. 1969. Topological models in biology. *Topology, 8,* 313–335.

Thom, R. 1970. Topological models in biology. (And

references therein.) In *Towards a theoretical biology: 3. Drafts, op. cit.*, pp. 86–116.

Thomas, H. A. 1971. Population dynamics of primitive societies. In S. F. Singer (ed.). *Is there an optimal level of population?* pp. 127–155. New York: McGraw-Hill.

Tischler, W. 1955. *Synokologie der Landtiere.* Stuttgart: Gustav Fischer Verlag.

Tischler, W. 1972. Discussion remarks at the XIV International Entomological Congress, Canberra.

Turnbull, A. L., and Chant, D. A. 1961. The practice and theory of biological control of insects in Canada. *Can. J. Zool., 39,* 697–753.

Ulanowicz, R. E. 1972. Mass and energy flow in closed ecosystems. *J. Theor. Biol., 34,* 239–253.

Vandermeer, J. H. 1970. The community matrix and the number of species in a community. *Amer. Natur., 104,* 73–83.

Vandermeer, J. H. 1972. On the covariance of the community matrix. *Ecology, 53,* 187–189.

Vandermeer, J. H. 1973. Generalized models of two species interactions: a graphical analysis. *Ecology,* in press.

van der Pol, B., and van der Mark, J. 1928. The heartbeat considered as a relaxation oscillation, and an electrical model of the heart. *Phil. Mag., 6,* 763–775.

Volterra, V. 1926. Variazioni e fluttuazioni del numero d'individui in specie animali conviventi. *Mem. Acad. Lincei., 2,* 31–113. (Translation in an appendix to Chapman's *Animal ecology,* New York, 1931.)

Volterra, V. 1931. *Leçons sur la théorie mathématique de la lutte pour la vie.* Paris: Gauthier-Villars.

Volterra, V. 1937. Principes de biologie mathématique. *Acta Biotheoretica, 3,* 1–36.

Wangersky, P. J., and Cunningham, W. J. 1957. Time lag in prey-predator population models. *Ecology, 38,* 136.

Watt, K. E. F. 1959. A mathematical model for the effect of densities of attacked and attacking species on the number attacked. *Canad. Ent., 91*, 129–144.

Watt, K. E. F. 1960. The effect of population density on fecundity of insects. *Canad. Ent., 92*, 674–695.

Watt, K. E. F. 1963. Dynamic programming, "Look ahead programming," and the strategy of insect pest control. *Canad. Ent., 95*, 525–536.

Watt, K. E. F. (ed.) 1966. *Systems analysis in ecology.* New York: Academic Press. See in particular Watt's own chapter, "The nature of systems analysis," pp. 1–14.

Watt, K. E. F. 1968. *Ecology and resource management.* New York: McGraw-Hill.

White, A., Handler, P., Smith, E. L., and Stetten, D. 1959. *Principles of biochemistry.* 2nd ed. New York: McGraw-Hill.

Whittaker, R. H. 1969. Evolution of diversity in plant communities. In *Diversity and stability in ecological systems, op. cit.,* pp. 178–196.

Williams, C. B. 1964. *Patterns in the balance of nature.* London: Academic Press.

Williamson, M. 1972. *The analysis of biological populations.* London: Edward Arnold.

Winsor, F., and Parry, M. 1958. *The space child's mother goose.* New York: Simon and Schuster. Poem 30.

Wright, S. 1938. The distribution of gene frequencies under irreversible mutation. *Proc. Nat. Acad. Sci., 24,* 253–259.

Zwolfer, H. 1963. Untersuchungen uber die struktur von parasitenkomplexen bei einigen lepidopteren. *Z. angew. Ent., 51,* 346–357.

Zwolfer, H. 1971. The structure and effect of parasite complexes attacking phytophagous host insect. In *Dynamics of populations, op. cit.,* pp. 405–418.

# Author Index

229

# AUTHOR INDEX

231

# Subject Index

233